So You Are a Chef

Managing Your Culinary Career

. . . .

Lisa M. Brefere
CEC, AAC

Karen Eich Drummond
Ed.D, RD, FADA, LDN, FMP

Brad Barnes
CMC, AAC

. . . .

WILEY

John Wiley & Sons, Inc.

Published by John Wiley & Sons, Inc., Hoboken, New Jersey.

Published simultaneously in Canada.

For general information on our other products and services, or technical support, please contact our Customer Care Department within the United States at 800-762-2974, outside the United States at 317-572-3993 or fax 317-572-4002.

Wiley also publishes its books in a variety of electronic formats. Some content that appears in print may not be available in electronic books.

For more information about Wiley products, visit our Web site at http://www.wiley.com.

Library of Congress Cataloging-in-Publication Data:

Brefere, Lisa M.
    So you are a chef : managing your culinary career / Lisa M. Brefere, Karen Eich Drummond, Brad Barnes.
        p. cm.
    Includes index.
    ISBN 978-0-470-25127-0 (paper/CD : alk. paper)
        1. Cooks—Vocational guidance.   2. Food service—Vocational guidance.   3. Cookery—Vocational guidance.
    I. Drummond, Karen Eich.   II. Barnes, Brad.   III. Title.
    TX649.A1B74 2009
    641.5029—dc22
                                                                                      2007044570

Printed in the United States of America
10 9 8 7 6 5

Contents

Introduction

You're about to complete your culinary education or are seeking a change of job. What now? Preparing a resume, putting together your portfolio, finding potential employers, interviewing, and generally advancing your career can be daunting. *So You Are a Chef* steps you through the process of all this in the hope of helping you achieve success.

We had originally combined the information in *So You Are a Chef* with the occupational path content of *So You Want To Be a Chef?* When it came time to update *So You Want To Be a Chef?* to a new edition, we looked carefully at the content and listened to our readers and decided to separate the material into two books. The result is *So You Are a Chef*, which is now a companion book to *So You Want to Be a Chef?, Second Edition.* Where *So You Want to Be a Chef?, Second Edition* gives you detailed information to direct your culinary career options, *So You Are a Chef* is the next step in managing your career, whether you are soon to graduate from a culinary program or have recently graduated, or if you have been working in the industry and are looking to make a change.

You have skills that employers want, but those skills won't get you a job if no one knows you have them. Good resumes, cover letters, applications, and a job portfolio all broadcast your abilities. *So You Are a Chef* walks you through the process to create these tools. They tell employers how your qualifications match a job's responsibilities. If these paper preliminaries are constructed well, you have a better chance of landing interviews—and, eventually, a job.

Technology has added a new twist to preparing resumes and cover letters and sending them to prospective employers. The availability of personal computers and laser printers has raised employers' expectations of the visual quality of resumes and cover letters applicants produce. The widespread use of the Internet to post jobs and resumes has also created situations where there are simply millions of resumes on the Internet, and a single job opening might generate hundreds or even thousands of responses. Busy reviewers often spend as little as 30 seconds deciding whether a resume deserves consideration. In some companies, if your resume is not formatted properly for computer scanning, it may never reach a human reviewer.

How to Use   *So You Are a Chef*

So You Are a Chef is organized into five chapters, as shown here:

Chapter 1 offers advice on how to write up your resume to help you pass the 30-second test and win interviews.

Chapter 2 helps you put together a job search portfolio that will highlight your accomplishments and make you stand out from other applicants.

Before you can interview, you must learn how to locate and contact prospective employers. Which of the following techniques do you think is most likely to get you a job?

1. The classified advertisements in the newspapers

2. The Internet

3. Networking contacts

The correct answer is #3. Networking contacts are much more likely than other sources to connect you with your future employer. **Chapter 3** helps you use a variety of techniques to find potential employers and write appropriate cover letters to send with your resume so that you get your foot in the door.

Chapter 4 discusses the steps involved in interviewing, including choosing what to wear and bring, anticipating questions you will likely face, closing the interview, and following up. If you prepare for the interview properly, as this chapter discusses in detail, you will certainly not be as nervous when the time comes, and you'll do a better job of projecting your knowledge, skills, and abilities.

The final chapter, **Chapter 5,** talks about your career. A career is not simply a series of jobs but rather a series of progressive achievements. This chapter discusses how you can set goals, work with others, involve mentors, be an active member in professional organizations, become certified, and participate in lifelong learning to ensure progressive achievements in a long, successful, and enjoyable career.

CD-ROM

This book is accompanied by a CD-ROM, which includes a resume worksheet and checklist, resume templates, a portfolio worksheet and checklist, a job log, templates for cover letters and thank-you letters, and a goal-setting chart.

Resources for Instructors

To assist the instructor, an online **Instructor's Manual** has been developed. It includes learning objectives, a chapter outline, class activities, and test questions for each chapter in the book. In addition, PowerPoint slides and student worksheets are available for downloading at www.wiley.com/college/brefere.

Acknowledgments

We also appreciate the feedback of the reviewers who read the initial proposal and subsequent drafts of the manuscript for *So You Are a Chef* and *So You Want To Be a Chef?, Second Edition,* and convinced us that we really had two books. They are:

John M. Courtney, Johnson County Community College

Emma Cutler-Grimes, New England Culinary Institute

Jamie M. Gizinski, Walnut Hill College

Sarah Gorham, The Art Institute of Atlanta

Joseph Renfroe, York Technical Institute

Odette Smith-Ransome, The Art Institute of Pittsburgh

David Weir, Orlando Culinary Academy/Le Cordon Bleu

Write Your Resume

Introduction

THE RESUME IS ONE OF THE MOST CRITICAL STEPS in securing a job. Remember that your resume precedes the interview and is the only impression you make on your potential employer before you are (hopefully) asked to make a personal impression in an interview. Most resumes are glanced at for less than a minute—less time than you might wait for a red light to turn green. Resumes that are wordy or hard to read end up in the trash, and many resumes wind up being filed away forever. Only a small percentage of resumes ever make it to the interview step.

But employers still ask for resumes, and a good resume provides a competitive edge. Your resume tells potential employers what you have accomplished already and what you can do for them now. Look at the resume as an advertising tool; it sells your talents and skills to an employer, much as a 60-second commercial sells to a consumer. Good resumes can awaken an employer's interest in you and get you what you want—an interview.

Types of Resumes

Resumes fall into one of three categories: chronological, functional, or combination (combines characteristics of both the chronological and functional). The type you choose should emphasize your strengths and deemphasize your weaknesses. Most resumes these days are combination resumes, as you will see shortly.

The chronological resume (Figure 1-1) lists the jobs you've had by date of employment, starting with your most recent job and working backward. The education section lists your education in reverse chronological order as well. Use the chronological resume if:

Figure 1-1
Chronological
Resume

◆ You have recent and continuous work history in the field you are looking for a job in.
◆ You have progressed up a clearly defined career ladder and are looking for advancement.

Cheryl Richardson

Permanent Address:
92 Longwood Road
Aurora, NY 11593
315-555-1212

cherylrich@yahoo.com

Current Address until June:
233 University Avenue
Ithaca, NY 12830
315-555-1213

Summary

Dean's List college student in culinary arts, recently promoted to Line Cook at nationally known Moosehead Restaurant.

Work Experience

7/08–Present **Line Cook** at Moosehead Restaurant, Ithaca, NY
Work at sauté or grill station for lunch or dinner meals in a well-known restaurant featuring healthful natural foods cuisine. Perform mise en place and food preparation. Follow safe and sanitary food procedures.
○ Test and evaluate new recipes.
○ Won Employee of the Month (June 2009).

10/06–6/08 **Preparation Cook** at Moosewood Restaurant, Ithaca, NY
Performed all preparation tasks in kitchen emphasizing scratch cooking and vegetarian dishes. Completed all duties in timely fashion while maintaining sanitation standards.
○ Received "Excellent" performance evaluations.

Summers 2005 and 2006 **Assistant Cook** at Lenape Summer Camp, Seneca Falls, NY
Under Head Cook's direction, did basic food preparation tasks, cooking, and baking. Assisted in purchasing, receiving, and inventory management.

Education and Certification

May 2009 Bachelor of Professional Studies in Culinary Arts, Olympia University, Ithaca, NY

Dean's List every semester (Anticipated)

Treasurer, Culinary Club (sophomore year)

ServSafe® Food Protection Manager, #2364656 (National Restaurant Association Educational Foundation)

Employers especially like to see a clearly defined career ladder in your listing of jobs; it lets them know what you can do right now. Do not use this type of resume if you are just starting out, trying to switch fields, or have large gaps in employment—then it would be better to use the functional resume.

The functional resume (Figure 1-2) also includes a listing of your work experience and education, but in a brief form toward the end of the resume. Most of the functional resume is a summary of your skills and accomplishments, such as

Figure 1-2
Functional
Resume

Tim Fitzpatrick

3626 Chestnut Drive, Sauna, CA 84529   408-392-8942   tfitz@aol.com

---

Seeking an entry-level Cook position in a restaurant.

College student in culinary arts with diverse cooking and foodservice experience, including food preparation and supervising.

Culinary Arts:
- Experienced with kitchen food preparation and cooking equipment.
- Competent in basic food preparation techniques, including cutting.
- Use standardized recipes.
- Follow portion control guidelines.
- Plate and garnish foods.

Sanitation
- ServSafe® certified.
- Follow appropriate cleaning and sanitation procedures.

Supervision
- Supervised five employees.
- Scheduled, trained, motivated, and coached employees.
- Solved problems.

Employment

Cold Food Preparation, Bay Community College.   September 2007 to present.
Part-time.

Head Waiter, The Tides Retirement Community.   June 2006 to August 2007.
Part-time.

Waiter, The Tides Retirement Community.   June 2005 to June 2006.
Part-time.

Education

Associate in Occupational Studies in Culinary Arts anticipated May 2008. Bay Community College.

specific culinary skills you've used or menus you've developed and served. Use the functional resume if:

◆ You are applying for a job that is quite different from your current or past job.
◆ You have little to no work experience in this field.
◆ You are reentering the job market after a break.

The functional resume emphasizes what you can do and deemphasizes where you have worked. Many skills, such as management skills, are transferable between industries, and this type of resume especially helps people who are switching to the culinary field or just starting out after college.

Many employers look on functional resumes with some level of distrust. While they can see what sorts of skills and abilities you have, they don't know where you learned them. This is a good reason to consider the next type of resume.

The combination resume (Figure 1-3) combines features from both the chronological and functional resumes into a type of resume that is increasingly popular. Basically, you showcase your skills and achievements at the beginning of the resume, typically in a section entitled Profile or Summary. Then you go on to describe your jobs and education in reverse chronological order. It's a format that almost any jobseeker can customize to meet his or her needs.

Figure 1-3
Combination
Resume

Richard Plumb, C.E.C., A.A.C.

211 West Greenwich Avenue
Greenwich, CT 07041
203-437-9365 (h) 203-530-8821 (c)
brewchef@yahoo.com

Profile

- Experienced Executive Chef and Director of Operations. Have operated multiple restaurants accommodating over 500 guests.
- Developed kitchen and menus for new brewery restaurants.
- Excel in developing successful menus and recipes.
- Proven team-building and motivational skills have kept staff turnover below 40%.

Experience

Director of Operations/Corporate Executive Chef
Boston Hops, Inc., New York, NY   May 2004–present
Responsible for menu development, kitchen/bar design, opening plan and execution, training, and staff hiring for three new brewery restaurants.

- Redesigned kitchen.
- Upgraded menus.
- Developed corporate buying policies, recipes, restaurant standards, and training manuals.

Executive Chef/Back of House Director of Operations
Greenwich Regency, Greenwich, CT   May 2000–May 2004
Responsible for 32 Cooks and 6 Sous Chefs in a $9.5 million
food and beverage operation. Also supervised stewarding,
purchasing, and receiving.

- Five-year average of 29% food cost and 30% labor cost.
- Employee retention improved 75%.
- Operation featured in numerous publications.

Chef de Cuisine
Pebble Creek Café, Purchase, NY   June 1997–March 2000
Instrumental in kitchen and restaurant design of American
regional restaurant. Responsible for all costs for front and back
of the house. Developed menus, monthly marketing tools, and
advertising strategies.

- Increased quarterly sales 25%.
- Demonstrated project planning and design skills.

Education

A.O.S. in Culinary Arts, April 1995
New Jersey Culinary Institute

Nutritional Cuisine, January 2004, New Jersey Culinary Institute,
20-hour course

Certification

Certified Executive Chef, January 2004
American Culinary Federation

Associations

Active Member, American Culinary Federation, since 1995

The Chefs Association of Westchester and Lower Connecticut
since 1995, President from 1998 to 1999

Awards

2002 Chef of the Year, The Chef's Association of Westchester and
Lower Connecticut

Delaware Valley Chefs Association Culinary Competition,
ACF Silver Medal, 2001

U.S. Team Member, International Ice Carving Competition,
Gold Medal, 2000

Southern New Jersey Chefs Association Culinary Competition,
First Prize, Poultry Platter, 1999

The Ingredients of a Great Resume

A great resume sells a potential employer the idea that you are the person to do the job. Your resume will do this most effectively if you remember that it is not just a job description of your current and past jobs. For a resume to be great, you need to:

◆ Choose and highlight the parts of your background that position you for the type of job you are currently seeking.
◆ Discuss what you did in other jobs, but especially how well you did it.
◆ Include measurable achievements and accomplishments.

Following are guidelines for what to include and what not to include on your resume.

WHAT YOU MUST INCLUDE

Most professional resume writers agree that you must include these sections in your resume.

◆ Contact information (including a businesslike email address)
◆ Profile (short summary of qualifications)
◆ Professional experience
◆ Education
◆ Professional licenses/certifications (such as ServSafe®)
◆ Professional affiliations (such as membership in the American Culinary Federation)

Additional sections that present information such as computer skills and awards are also appropriate.

WHAT YOU MIGHT INCLUDE

You might include a job objective, a short statement of the type of job you are looking for. It is important that the job objective be concise and not too broad—for example, "Job Objective: Sous Chef in Club Setting." Some applicants like to use an objective; others don't. The information stated in your objective will be stated in your cover letter, so it is not absolutely essential that it be on your resume. However, if you are not sending your resume in for a specific job opening, it's a good idea to include a job objective because the employer is not immediately associating your resume with a specific opening.

Place your job objective below the contact information on the resume and check that it is appropriate each time you send your resume out. You want your stated job objective to closely match the job you are applying for.

WHAT TO OMIT

Don't put any of these on your resume:

◆ Reference information (just state that a list of references is available)
◆ Availability
◆ Salary history
◆ Diversity issues
◆ Photographs

It is customary to give out your reference list only at an interview or after you have been interviewed for a job. Availability is also a subject that can be addressed in an interview. You don't want to advertise that you are available immediately—it makes you look desperate! Salary is yet another issue that should be discussed later. As described in the chapter on interviewing, it is best not to discuss salary with the employer until you are offered the job. Once you receive an offer, you are in a much better position to negotiate a good salary.

How to Write Your Resume

The type of resume discussed in detail here is a combination resume, which begins with a profile in which you highlight your qualifications and accomplishments. Then it moves on to a chronological review of your professional (work) experience, education and certifications, professional affiliations, and other information you want to include.

CONTACT INFORMATION

At the top of every resume is your contact information, including your mailing address, telephone numbers, and email address. It is acceptable to use the postal abbreviation for your state instead of writing out the name of your state. For example, use CT for Connecticut. Don't use any abbreviations for your street address (such as Ave. for Avenue) or city (such as NYC for New York City).

When typing out your phone number(s), be sure to include your area code and designate which number is which, as shown in the following example.

(H) 272-356-7890

(C) 272-367-5237

You may put parentheses around the area code, but don't put 1 before the area code. Make sure you have a reliable answering service for every phone number you put on your resume, including cell phones. Of course, once you send your resume to potential employers, you must frequently check for voicemail messages.

If you are a student still in college, it is best to give both your college and home addresses and telephone numbers and to note when to use each address. For example, you might state next to your college address something like "Contact through May."

Email represents yet another way to communicate with employers. You should definitely list an email address, one that sounds professional. Don't type partyguy@aol.com on your resume and then wonder why you aren't getting any phone calls. Job hunting requires a suitably professional email address. Your Internet provider may allow you to pick several email addresses, so choose one with a neutral feel. It's quite common for job hunters to reserve one email address for the resume. If you want a new email address, check out the free email accounts available from companies such as hotmail.com.

PROFILE

Your profile section appears right below your contact information so the employer can quickly get an idea of who you are, what you can do, and how you can contribute. This section can be titled Profile, any of the following names, or any appropriate name you can think of.

- Career Profile
- Professional Profile
- Summary
- Qualifications
- Summary of Qualifications
- Areas of Expertise (or Proficiency)
- Key Strengths
- Core Competencies
- Professional Highlights
- Achievements or Accomplishments
- Highlights of Skills and Experience
- Highlights of Qualifications

This section can take the form of a bulleted list, a paragraph, or both. Whichever format you choose, make this section brief and focused. Highlight your experience, accomplishments, and skills. As you write a rough draft of this section, make sure it answers this question: "If I had only 30 seconds to get someone to hire me, what would I say?"

When you mention your skills, be sure they are directly related to the type of position you want. Also, stating "hardworking employee" is not nearly as strong as evidence such as "Promoted from preparation cook to line cook within three months because of excellent knife skills and work ethic."

Here is a bulleted profile for a highly experienced Certified Master Chef.

Profile

- Successfully completed Certified Master Chef test.
- Over 20 years' experience in quality food preparation.
- Thorough understanding of all facets and styles of foodservice.
- Well versed in many ethnic and international cuisines.

- Able to produce quality results while adhering to well-planned budgets.
- Over 8 years' experience in multi-unit management.
- Excellent human resource management skills, maintaining a departmental employee retention average of 3.5 years.
- Self-motivated quality- and cost-oriented manager.
- Highly trained in nutritionally conscious cuisine.

Of course, this profile is pretty long because of the chef's extensive and noteworthy culinary career. Yours will most likely have fewer bullets. Note that the most significant achievements are noted first. The format of this profile could be changed by combining the first three bullets into a short paragraph and then bulleting the remaining points.

A profile for someone coming out of college with some work experience in the industry might look like this.

Profile

Hardworking and reliable culinary student distinguished by:

- Over two years' experience as a preparation cook promoted to line cook at La Brasserie.
- Silver Medal earned in ACF-sanctioned hot food competition, category K.
- President's Honor Roll every semester.
- Strong interpersonal and organizational skills.

Make sure that some of the points you make are measurable achievements, like earning a silver medal at a culinary competition, saving the department $25,000 a year in labor, and cutting staff turnover by 25%.

You may want to write up your profile after you have completed the work experience and education sections. Once you have those sections ready, it will be easier for you to see which skills, achievements, and experience you want to highlight in the profile.

PROFESSIONAL EXPERIENCE

After the profile, your next section will probably be professional experience, although in some cases it may be education. A good rule of thumb is to put the stronger section first. For example, if you are seeking a job as a culinary educator, the amount of formal education you have is important, so you may want to highlight your degrees at the top of the resume. For most culinary positions, put your work experience first unless you have almost no experience.

You don't have to call this section Professional Experience. Other possible names include the following.

- Professional Background
- Employment History
- Work History
- Work Experience
- Experience

- Career Track
- Employment Chronicle
- Career History
- Career Path

To start writing up your work experience, use the Resume Worksheet to write down information about your jobs. Start with your current or most recent job, and then work backward. The worksheet will help you decide what to include on the resume.

When writing up your job duties, think in terms of the broad responsibilities you had and the specific tasks or duties you performed for each responsibility. For example, on your resume you could state the broad responsibility first, and then present a bullet list of important duties and notable accomplishments. Don't just discuss what you did; also include how well you did it. Employers want to see measurable achievements. It helps them see how you can contribute to their organization's bottom line. Here's an example.

EXECUTIVE CHEF June 2005–July 2009

Big Oak Café, Troy, New York

Supervised and coordinated the food purchasing and production for kitchen producing 1,000 meals/day.

- Purchased over $1 million of food and supplies yearly.
- Saved $25,000 in the first year after improving bid system and updating purchase specifications.
- Developed and instituted regular seasonal menu changes.
- Reduced kitchen labor cost by 5%.
- Quality of food consistently rated "good" or higher.
- Conducted formal monthly training sessions and daily coaching of employees.

Note three points in this example. First, there are no complete sentences; each statement is a phrase. Second, each phrase begins with a specific, descriptive verb. For example, instead of a general verb such as manage, use precise verbs such as organize or direct. Table 1-1 lists action verbs you can use when preparing your resume. Try to avoid phrases that begin with "Responsible for"; instead, find an appropriate verb. Third, once you have climbed the career ladder, it is assumed you can cook. So talk about how many people you supervised, the volume of the business, and how you managed costs. If your experience is mostly cooking, be careful of repetitive wording when describing your jobs. For example, don't keep listing "sautéed fish and chicken" for each job.

Table 1-1 Verbs for Resumes

Communication Skills

arranged	composed	edited	motivated	publicized
addressed	conferred	explained	negotiated	published
authored	corresponded	formulated	persuaded	wrote
clarified	drafted	informed	presented	

Creative Skills

conceptualized	designed	fashioned	illustrated	originated
created	established	focused	invented	performed

Culinary Skills

arranged	converted	filleted	judged	prepared	simmered
assembled	cooked	finished	measured	produced	specified
baked	cooled	flavored	microwaved	purchased	steamed
boiled	cut	formulated	pan-broiled	roasted	stored
braised	deep-fried	garnished	pan-fried	sautéed	thickened
broiled	designed	griddled	performed	scaled	used
calculated	determined	grilled	planned	seasoned	
chose	dressed	identified	poached	set up	

Financial Skills

administered	computed	formulated	purchased	sold
analyzed	contracted	increased	recommended	trimmed
audited	cut	marketed	reconciled	
balanced	decreased	planned	recorded	
budgeted	eliminated	projected	reduced	
calculated	forecast	provided	saved	

Human Resource Skills

coached	encouraged	instructed	oriented	specified
counseled	evaluated	interviewed	placed	staffed
delegated	facilitated	mediated	promoted	streamlined
developed	guided	moderated	recruited	taught
empowered	helped	motivated	represented	trained
enabled	hired	negotiated	screened	

Table 1-1 Verbs for Resumes (continued)

Management Skills

accepted	created	finished	optimized	restored
accomplished	defined	focused	organized	restructured
achieved	delivered	founded	originated	revamped
adapted	demonstrated	formulated	overhauled	revitalized
administered	designed	generated	oversaw	saved
advanced	developed	guided	performed	scheduled
advised	devised	headed	persuaded	solved
allocated	diagnosed	identified	planned	spearheaded
analyzed	directed	implemented	prepared	streamlined
appraised	diversified	improved	presented	structured
approved	eliminated	increased	presided	summarized
assigned	engineered	innovated	prioritized	supervised
assisted	enlisted	inspected	processed	surveyed
chaired	established	installed	produced	traveled
clarified	evaluated	instituted	provided	trimmed
conducted	examined	introduced	regulated	upgraded
consolidated	executed	launched	remodeled	
contributed	expanded	led	repaired	
controlled	expedited	maintained	represented	
coordinated	facilitated	monitored	resolved	

Marketing and Sales Skills

compiled	distributed	generated	maintained	obtained
consolidated	expedited	increased	marketed	stimulated

Another way to format your work experience is to start with a short paragraph listing your responsibilities and duties, and then have a bulleted list of your accomplishments. Here is how that approach looks:

EXECUTIVE CHEF June 2005–July 2009

Big Oak Café, Troy, New York

Supervised and coordinated the food purchasing and production for kitchen producing 1,000 meals/day. Purchased over $1 million of food and supplies yearly. Developed and instituted regular seasonal menu changes. Conducted formal monthly training sessions and daily coaching of employees.

Performance Highlights

◆ Saved $25,000/year after improving bid system and updating purchase specifications.

◆ Reduced kitchen labor cost by 5%.

◆ Quality of food consistently rated "good" or higher.

When thinking of your accomplishments and achievements, ask yourself if you ever did the following:

1. Save your employer money—if so, how much?
2. Increase sales—if so, how much?
3. Increase profitability—if so, how much?
4. Bring in new business—if so, how much?
5. Increase employee retention—if so, how much?
6. Decrease payroll costs, including overtime—if so, how much?
7. Increase guest satisfaction—if so, how much?
8. Increase profitability—if so, how much?
9. Decrease or keep food cost constant—if so, how much?
10. Increase check average—if so, how much?
11. Reduce purchasing costs—if so, how much?
12. Update and improve policies and procedures
13. Initiate and implement new menus or programs
14. Implement new hardware, software, or other systems
15. Improve productivity
16. Improve communications
17. Design new training programs
18. Introduce new standards
19. Streamline operations, functions, or support activities
20. Realign staffing to meet business demand and/or decrease costs
21. Receive a prize/honor/award from an employer, school, or professional organization
22. Manage special projects, such as kitchen renovation or purchasing new equipment
23. Develop unique skills or qualifications
24. Have public speaking experience
25. Have culinary industry certifications

Quantify your achievement whenever possible, as in "Increased check average 5%."

EDUCATION AND CERTIFICATIONS

Next, discuss your education and certifications. Conceptually divide this section into three parts:

1. College
2. Continuing education (or lifelong learning)
3. Certifications

If this section is long, you can certainly separate it into two or three sections. As long as you are in college or have graduated from college, you probably do not need to include high school information. If you went to a particularly prestigious high school or one with a well-known culinary program you were in, you might include the name of the school and program and the year you graduated.

With regard to your college education, the following items are the bare minimum you must put on your resume.

- Type of degree received, your major, and date of graduation—always list the degree before the name of the college or university at which you earned it. If you have graduated, you could use this format:

  Bachelor of Professional Studies in Culinary Arts, 2008

  Culinary University, Denver, Colorado

  If you are still in school, give the month and year when you anticipate completing your degree. For example:

  Associate in Occupational Studies in Culinary Arts anticipated May 2010

  Culinary University, Denver, Colorado

  If you are not that close to finishing your degree, you could say this:

  Currently pursuing an associate degree in Occupational Studies in Culinary Arts

  Culinary University, Denver, Colorado

  If you minored at college in an area related to culinary, mention that as well. If your college major was unrelated to the culinary field (such as German or history), mention your degree but don't specify your major.

- Names of colleges and universities you've attended—if you transferred from a community college, for example, to a four-year college and earned your degree, it is not absolutely essential to mention the community college. However, if it might work to your benefit to mention the community college, as when the community college's culinary program is well known, include it on your resume. You can also include your cumulative average if it is good—meaning at least over 3.0 if your school uses a standard 4.0 scale. List your cumulative average like this: 3.0/4.0.

Of course, you can include many other aspects of your college education on your resume.

- **Academic honors**—Note academic honors such as Dean's List, awards, honor societies, and scholarships.
- **Internships**—Mention where you completed your internships; note the time frame and what you did.
- **Activities**—Many college students don't have much work experience, so listing involvement in school or extracurricular activities is important. Employers look for this because such involvement shows initiative. If you were involved in a culinary club or association, especially if you held an

How to Write Your Resume

office, include this information on your resume. Holding an office shows leadership. Include volunteer activities.

- **International study**—Include where you studied, when, and a brief statement of what you did.
- **Special projects/Team projects**—If you don't have much work experience, you may want to briefly describe a special college project, perhaps a team project, if it is related to the position you are applying for. For example, you may have worked on a project involved in catering events or culinary competitions.
- **Courses taken**—Listing four to eight relevant courses may benefit you if you are a recent graduate and don't have much work experience.

After your college section, mention relevant continuing education courses you have taken. These could include classes provided by an employer, workshops or seminars attended at industry-related conferences, continuing education courses taken to maintain American Culinary Federation certification, and formal education courses such as computer classes taken online or in the classroom. Specify the year in which you took the training. If the training was particularly lengthy, you can also add the number of hours or days it required. Don't forget to include computer courses you have taken.

You can also include certifications, such as Certified Culinarian, in this section, or you may want to list them in a separate section. Specify the certification you have, the certifying organization, and when you received the certification.

PROFESSIONAL AFFILIATIONS

Your memberships in appropriate professional associations show your enthusiasm and dedication to your career. Membership is also important for keeping up in the field and networking with colleagues.

ADDITIONAL SECTIONS

- **Computer skills**—Every job requires computer skills. List the software programs you can use with at least basic proficiency.
- **Foreign language skills**—If you are fluent in a language other than English, especially Spanish, do mention it on your resume. If you are not fluent but can read, write, or speak well, include this information too. Just make sure you write down, for example, "Speak Spanish" or "Read French."
- **Volunteer work**—Chefs frequently do volunteer work with food banks and other organizations. Mention relevant volunteer work you have performed, along with the name of the organization and the year.
- **Awards/Honors**—List awards and honors you received, from employer awards to medals won at culinary competitions. Give the name and year and describe the award/honor, if necessary.
- **Military service**—Mention the branch of service in which you served, your highest rank, your dates of service, decorations or awards, and special skills or training you received that could further your career.

◆ **Publications**—If you have published an article in an industry magazine, a book, or any other relevant material, list the title and publication date.

◆ **Presentations**—List presentations you made at professional meetings and in other professional settings.

REFERENCES

Resumes usually do not list names of references. Most resumes close with the statement "References available on request."

RESUME WORKSHEET

Figure 1-4
Resume Worksheet

Use the Resume Worksheet (Figure 1-4, also on the CD-ROM) to help organize the information for your resume.

CONTACT INFORMATION

Name: _____

Home Address: _____

_____

School Address (if applicable): _____

_____

Telephone Numbers: _____

_____

Email Address: _____

Other Information: _____

PROFILE

_____

_____

_____

_____

_____

_____

PROFILE

_____

_____

_____

_____

_____

_____

PROFESSIONAL EXPERIENCE

Dates Employed (month/year): _____

Employer's Name: _____

Employer's Address (City/State): _____

Job Title: _____

Responsibility #1 _____

    Duties and Accomplishments

    _____

    _____

    _____

    _____

Responsibility #2 _____

    Duties and Accomplishments

    _____

    _____

    _____

    _____

Responsibility #3 _____

    Duties and Accomplishments

    _____

    _____

_____

_____

Responsibility #4 _____

    Duties and Accomplishments

    _____

    _____

    _____

    _____

EDUCATION

School: _____

Major(s): _____

Minor: _____

Date Graduated/Dates Attended: _____ Degree Seeking/Granted: _____

Grade Point Average: _____

Academic Honors _____

_____

_____

_____

Scholarships: _____

Co-ops or Internships (Where, When, Description)

_____

_____

_____

Extracurricular Activities: _____

_____

_____

_____

Offices Held: _____

_____

International Travel (Where, When, Description):

_____

_____

_____

Special Projects/Team Projects (When and Description): _____

_____

_____

Relevant Coursework (only for current students and recent graduates):

_____

_____

_____

_____

_____

High School (if going to include): _____

Continuing Education: _____

_____

_____

_____

_____

Certifications (certificate number and expiration date when applicable):

_____

_____

_____

_____

PROFESSIONAL AFFILIATIONS

_____

_____

_____

_____

OPTIONAL SECTIONS

Volunteer Work:

_____

_____

Computer Skills:

_____

_____

Foreign Language Skills:

_____

Awards/Honors:

_____

_____

Publications:

_____

_____

Presentations:

_____

_____

MAKING A ROUGH DRAFT

Once that you have completed the Resume Worksheet, you are ready to make a rough draft of your resume. For your rough draft, concern yourself with the information you want to include and how you want to say it. At this stage, you just want to decide what to say and what to leave out. Use Figure 1-5, which is also on the CD-ROM, to start working on your rough draft.

Don't plan to tackle this project in one night. You will need a number of work sessions to get a rough draft that you can type up. Take time to edit your resume at every step. Ask friends, teachers, and family for ideas and feedback.

Use these guidelines while preparing your rough draft.

Contact Information _____

Profile (or other name) _____

Professional Experience (or other name) _____

Education and Certifications (or other name) _____

Professional Affiliations (or other name) _____

Additional Sections (Volunteer Work, Computer Skills, Foreign Language Skills, Awards/Honors, Publications, Presentations)

Figure 1-5 Resume Draft

LENGTH

The length of your resume usually depends on the amount of work experience you have. Although you may have been told that your resume shouldn't exceed one page, if you have carefully chosen relevant material that requires two pages, that's fine. Resumes are frequently two and sometimes three pages long. If you have over ten years of experience, a two-page resume is common. Make each page a full page. If your last page has just a few lines on it, compress your information to remove the excess page.

FONTS AND FORMATTING

Use the following guidelines to format your resume.

- As for any business document, allow 1-inch margins on the sides, top, and bottom of each page.
- Double-space between sections and entries. Single-space paragraphs and bulleted lists. Be consistent with your spacing.
- Paragraphs should be short—five or six sentences at most. Break longer paragraphs into two or more.
- Pick out a font that is up-to-date and crisp.
  - Arial
  - Bookman
  - Century Schoolbook
  - **Franklin Gothic**
  - Garamond
  - Palatino
  - Tahoma

Times New Roman is also an acceptable font, but because it is used frequently, it is less distinctive than those noted here. The serif fonts—those where the letters have small lines extending from them, usually at the top and bottom—are often easier to read than sans-serif fonts. Examples of serif fonts include Bookman, Century Schoolbook, Garamond, and Palatino. Arial, Franklin Gothic, and Tahoma are sans serif fonts. Avoid fonts such as Courier that give the same amount of space for each letter, even though some letters are wider than others.

- In most cases, your best font size will be 10, 11, or 12, although headings and your name should be taller. For example, if you use 12-point Arial for the body text, try 14-point Arial for section headings and 16- to 18-point Arial for your name at the top of the resume.
- Don't clutter your resume with too much text. Lots of white space makes your resume easier to read. White space is the space on a page not occupied by text or pictures.
- Use underlining sparingly, if at all. Instead of underlining, try boldface, which is often better at getting attention. Further, it's easier to read boldface type than underlined text. Boldface works well for section titles and job titles. Don't use boldface to attract attention to a word or phrase in a sentence.

◆ Don't type words with all capital letters; this is very difficult to read. Instead, capitalize the first letter (if appropriate) and then switch to lower case.

◆ When you make a list, use a bullet (round, square, or diamond-shaped) or a tiny box instead of a hyphen. Use the same bullet style for each section or for the entire resume.

◆ Use a horizontal line to separate your contact information from the rest of the resume. The line helps organize the contents of the resume. You might set off each section with horizontal lines, as in Figure 1-1.

◆ To give your resume a consistent flow, maintain the same style from beginning to end. Every section should have the same design elements. For example, if your education heading is bold and centered, every section heading should be bold and centered.

FORMATS

Most resumes are set in either one or two columns. The resumes in Figures 1-2, 1-6, and 1-7 use one column, while the resumes in Figures 1-1 and 1-3 use two columns. The one-column format allows you to fit a little more information on a page because more space is available (but you still must leave plenty of white space). You can certainly type up your resume in both formats and then decide which looks best. You can even combine both formats by using one column for your contact, objective, and profile sections and then switching to a two-column format for the rest of the resume.

Let's take a look at the five resumes in this chapter to develop a better idea of ways to format a resume.

◆ Figure 1-1. The body of this resume is set in 12-point Arial, the section heads in 14-point bold, and the person's name in 16-point bold. In this use of the two-column format, the dates of employment and college are placed in the left column, and the job and education information are in the right column. The horizontal line under each section heading, along with the appropriate use of white space, make the headings stand out and result in an easy-to-read resume.

◆ Figure 1-2. This one-column resume uses 12-point New Century Schoolbook as the body font. The section names and contact information are set in 14-point type and the person's name in 16-point. The name, job titles, and section names are bolded. Because this is a functional resume, the middle has horizontal lines that emphasize the person's skills. Italics are used in this section for the top line.

◆ Figure 1-3. This two-column resume uses a T set of lines to make it look appealing. The font is Palatino, with 12-point type for the body, 16-point for the section heads, and 18-point for the person's name. The name, section heads, and job titles are bolded. The first column contains the section names and the second column the dates and information. If you like how this format looks but have more than one address to list, you can start your vertical line just below the contact section.

◆ Figure 1-6. This one-column resume shows the section heads in italic bold and centered with horizontal lines above and below for emphasis. The body

Cheryl Richardson

Permanent Address:
92 Longwood Road
Aurora, NY 11593
315-593-8270

cherylrich@yahoo.com

Current Address until June:
233 University Avenue
Ithaca, NY 12830
315-229-5987

Summary

Dean's List college student in culinary arts, recently promoted to Line Cook at nationally known Moosehead Restaurant.

Work Experience

Line Cook at Moosehead Restaurant, Ithaca, NY   July 2008–Present
Work at sauté or grill station for lunch or dinner meals in a well-known restaurant featuring healthful natural foods cuisine. Perform mise en place and food preparation. Follow safe and sanitary food procedures.
● Test and evaluate new recipes.
● Employee of the Month (June 2009).

Preparation Cook at Moosewood Restaurant, Ithaca, NY   October 2006–June 2008
Performed all preparation tasks in kitchen emphasizing scratch cooking and vegetarian dishes. Completed all duties in timely fashion while maintaining sanitation standards.
● Received "Excellent" performance evaluations.

Assistant Cook at Lenape Summer Camp, Seneca Falls, NY   Summers 2005 and 2006
Under Head Cook's direction, did basic food preparation tasks, cooking, and baking. Assisted in purchasing, receiving, and inventory management.

Education and Certification

Bachelor of Professional Studies in Culinary Arts   Anticipated May 2009
Olympia University, Ithaca, NY

Dean's List every semester

Treasurer, Culinary Club (sophomore year)

ServSafe® Food Protection Manager, #2364656 (National Restaurant Association)

Figure 1-6   One-Column Resume Sample

Brad Barnes, C.M.C., C.C.A., A.A.C.

213 Davis Avenue, Christianson, NY 10735
203.555.0150
BandBsolutions@aol.com

Qualifications

Culinary Skills

- Very strong experience in quality food preparation.
- Thorough understanding of all facets and styles of service.
- Well versed in many ethnic and international cuisines.
- Able to produce quality results while adhering to budget.
- Committed to upholding the highest standards of operation in the professional kitchen.
- Highly trained in nutritionally conscious cuisine.

Management Skills

- Self-motivated, quality- and cost-directed manager.
- Solid experience in multi-unit management.
- Excellent human resource management skills, maintaining a departmental employee retention average of 3.5 years.
- Skilled in sanitary management of food preparation facilities.
- Experienced in public speaking, presentations, and seminars.
- Developed, wrote, and presented educational videos.

Professional Experience

Chef/Owner, B & B Solutions   2001–present

Partner in Food and Beverage Management firm currently operating food and beverages operations in two Manhattan properties: the Embassy Suites Hotel in Battery Park City and the Hilton Times Square. Food and Beverages is a freestanding entity and is required to be totally self-sustaining while providing 24-hour room service, an employee cafeteria, and many other hotel services.

- Report on profitability, quality, and operations to our client.
- Reversed the operations from substantial losses to break even.

Corporate Executive Chef, ITB Restaurant Group   1992–2001

Oversaw profitability, training, menu development, and staffing of kitchens in three restaurants while acting as executive chef for the flagship operation, 64 Greenwich Avenue.

Figure 1-7   Resume of Very Experienced Chef

64 Greenwich Avenue Restaurant, 125 seats/$2.4 million annual sales
Responsible for design of the kitchen as well as the purchase of all equipment. Developed all menus. Developed profit and loss prospectus for opening food sales.

- Increased profitability of food sales by 10% since the opening through a customer-driven sales-oriented approach to menu development as well as a concentrated effort to retain employees and increase productivity.
- Practiced an aggressive approach to purchasing by constantly researching new resources while maintaining a good business relationship with purveyors.
- Initiated our banquet/catering division in order to expand sales as well as make better use of available staff and facility.
- Banquet/catering has grown to 35% of annual sales at a higher profitability than à la carte service.
- Maintained a constant learning atmosphere in the kitchen through promotion from within and the rotation of culinary school externs in the facility.

The Black Bass Grille, 65 seats/$1.4 million annual sales
The Black Goose Grille, 120 seats/$2 million annual sales
Set tone and style of menus and worked with the Chef to produce profitable, customer-driven menus that stayed within our philosophy and food standards. Wrote and implemented all front-of-the-house training procedures.

- Assured profitability of each kitchen through guidance in food cost control, staffing, and time management.
- Produced all graphics for seasonal menus.

Executive Chef, The Black Bass Grille   1989–1992
Hired to change the style of food and service from a tavern-style pub to a white-tablecloth casual dining restaurant.

- Raised check average from $20 to $37.
- Increased yearly sales from $780,000 to $1.4 million.
- Analyzed lunch business, which showed a history of poor customer counts, then recommended closing for that meal period, saving the company about $16,000 annually.
- Purchased new equipment per budget to facilitate new style of service.

Executive Chef, Greenwich Island Catering, $1.8 million annual sales 1987–1989
Supervised all food production and event logistics, including staffing and equipment setup. Maintained three daily corporate accounts.

Executive Chef, The Brass Register at Four Squares, 225 seats, 240 banquet seats, $1.6 million annual sales   1980–1985

Worked as Sous Chef and then Executive Chef. Started catering and banquet service.

Education and Certifications

A.O.S. in Culinary Arts, Culinary Institute of America, 1987

Nutritional Cuisine course, Culinary Institute of America, 1995

Certified Master Chef, American Culinary Federation

Certified Culinary Administrator, American Culinary Federation

Certified ACF International Judge

Certified ServSafe® Food Protection Manager

Certified TIPS Alcohol Service Trainer

Professional Organizations

Member, American Culinary Federation

Member, American Academy of Chefs

Member, World Association of Master Chefs

Honors and Awards

President's Medal from the American Culinary Federation

Coach and Design Director for American Culinary Federation Team USA, 2004 and 2000

Hermann Rusch Humanitarian Award for Contributions to 9/11 Relief Effort

Two Gold Medals, IKA/HOGA Culinary Olympics, Frankfurt, Germany

"Chef of the Year," The Chefs Association of Westchester and Lower Connecticut

font is 12-point Garamond, and the name font is 18-point. Section heads, job titles, and the person's name are bold. The years of employment are kept to the right.

♦ Figure 1-7. This resume is typical of someone with a lot of experience, expertise, and involvement in the culinary profession. The font is Franklin Gothic (12-point body text, 14-point section heads, and 16-point name). Section names, job titles, and employer names are bolded. The section headings appear to the left and have a horizontal line coming out to add emphasis and clarity. The body text is tabbed in to make the section headings more prominent.

The formats of these resumes can also be found on the CD-ROM.

KEYWORDS

Keywords are nouns or noun phrases that state job titles, skills, duties, and accomplishments (see Table 1-2). Some employers scan resumes into a database.

Table 1-2 Culinary and Management Keywords

Culinary Keywords

Back-of-the-house operation	Food service management	Mise en place
Banquet operations	Front-of-the-house operations	Multi-unit operations
Banquet sales	Garnish	Profit and loss responsibility
Budget administration	Guest relations	Portion control
Catering operations	Guest satisfaction	Presentation
Club management	Information technology	Product positioning
Corporate dining room	Inventory control	Project design
Customer retention	Labor cost controls	Project management
Customer service	Leadership	Purchasing
Employee training	Marketing	Sales
Food and beverage operations	Menu planning	Service management
Food cost controls	Menu pricing	

Management Keywords

Benchmarking	Leadership	Problem solving
Communication	Leadership development	Profit and loss management
Consensus building	Long-range planning	Quality improvement
Corporate culture	Multi-unit operations management	Sales management
Corporate mission	New business development	Team-building
Decision making	Organizational development	

Keywords help the employer identify applicants who may be able to fill a specific position. This is described in length in a moment. For now, you want to use appropriate keywords when possible in your resume. Another source of keywords is job advertisements.

VOICE AND TENSE

Even though you never say I on a resume, the subject of each phrase is indeed I. Be sure your verbs agree with the first person. Use the past tense of verbs when talking about past jobs and events. Use the present tense when describing what you do in your current job.

SPELLING, PUNCTUATION, AND GRAMMAR

When in doubt, use a good dictionary and a style guide. Use the dictionary to determine when certain words are hyphenated or capitalized. Also:

- Capitalize job titles, department name, company name, and towns/cities. Capitalize the first word of each bulleted item.
- Do not use abbreviations. Spell out abbreviations and acronyms, unless they are certifications that follow your name. For example, in Ron Smith, CPC, Certified Pastry Culinarian does not need to be spelled out.
- It is common practice to spell out numbers one through nine and then write the numbers 10 and above as numerals.
- Use colons and semicolons correctly, as well as apostrophes. Remember that it's means "it is," and the form its' does not exist in English.
- Put one space between a period and the first letter of the next sentence.
- Put a comma between a job title, the company name, and the location.
- Always put a comma between the name of a town or city and the state.

HONESTY

This guideline is simple: Be honest. Don't even try to be dishonest. The culinary world is really quite small, and you don't want to get a reputation for twisting facts. Even if you get something past an employer who hires you, many contracts include a clause that says dishonesty in the hiring process can result in job termination later.

PAPER

As you can guess, white or conservative colors such as ivory and light gray are best for resumes.

If you use watermarked paper, be sure to print your resume on the correct side of the paper. Hold up a piece of watermarked paper to the light; the correct side is facing you if you can read the watermark. Be sure the paper you use is at least 20# weight and is suitable for your printer.

Edit and Proofread

Once you have typed up a resume, it's time again to edit and proofread. The most common mistakes are simple typographical and spelling errors. Computer spellcheckers do not catch correctly spelled words used incorrectly—of for on, for example, or their for there. You want your resume to stand out, but not for the wrong reasons. Avoid mistakes: Have several people proofread your resume before you send it anywhere.

Use Table 1-3, Resume Checklist, to make sure you have a polished product.

Table 1-3 Resume Checklist

_____ 1. Is your resume easy to read?

_____ 2. Is your resume attractive?

_____ 3. Is there enough white space? Is each section distinct?

_____ 4. Have you kept every paragraph under five lines?

_____ 5. Is your contact information all correct?

_____ 6. Are your qualifications at the top of the resume easy to scan? Do they make you an attractive candidate? Does the list include at least one substantial accomplishment?

_____ 7. Does your resume highlight relevant education and work experience?

_____ 8. Does your work experience include measurable accomplishments?

_____ 9. Did you use action verbs when describing past work experiences?

_____10. Have you omitted references to salary and reasons for leaving jobs?

_____11. Is your highest educational attainment shown first?

_____12. Have you included relevant continuing education?

_____13. Have you included certifications you have, such as sanitation?

_____14. Did you mention special work-related skills?

_____15. If you are still in college, did you mention college activities and clubs you were involved in and offices you held?

_____16. Have you proofread your resume and allowed at least one other person to edit and proofread as well?

_____17. Can someone quickly glance at your resume and see the most important points?

Scannable Resumes

Many large companies, and a growing number of small ones, use computers to sort the hundreds of resumes they receive. These companies scan paper resumes into a computer database. When managers need to fill a position, they program the computer with keywords that describe the qualifications they want in a candidate. The computer then searches its database for resumes that include those keywords. The resumes with the most matches are forwarded to the managers.

Before you submit your resume to a company, call them to find out if it scans. If it does, be sure your resume's design is computer-friendly. Resumes that will be scanned should contain no graphics or formatting that a computer might misinterpret. Follow these steps to increase a scanner's ability to read your resume:

- Use nontextured white or very light paper with black letters.
- Choose a plain, well-known font such as Helvetica, Arial, or Times New Roman.
- Use a 12-point font for all body text and 14-point for all headings.
- Do not use underlines or italics, and do not use asterisks or parentheses. Modern systems can understand bold, but older systems might not. You can still distinguish headings by using capital letters.
- Use a one-column format.
- Avoid boxes, graphics, columns, and horizontal or vertical lines.
- Put your name on its own line at the top of each page. Also, give each piece of your contact information (address, phone number, email address) its own line.
- Use round, solid bullets.
- Do not staple or fold your resume.

Figure 1-8 contains an example of a scannable resume.

Everything You Need to Know About References

Before making a hiring decision, most employers want to speak with people who know you well. You should find three to five people who agree to recommend you to potential employers.

Choosing references can be difficult, especially for people with little work experience. But you may have more options than you think. The people you ask to be references should be familiar with your abilities. Supervisors from paid or unpaid jobs, teachers, advisors, coaches, and coworkers are all good choices. Select the most willing, articulate people you can. Always ask permission of the people you want to speak for you before including their name on your reference list.

After choosing and contacting references, type a list with the name, address, telephone number, and email address of each one, and briefly describe his or her relationship to you. Bring copies of this list, along with your resume, to interviews.

When people agree to be references, help them help you. Send them a copy of your resume or application to remind them of your important accomplishments. Tell them what kinds of jobs you are applying for so they know what types of questions to expect.

Cheryl Richardson

92 Longwood Road
Aurora, NY 11593
315-593-8270
cherylrich@yahoo.com

Summary

Dean's List college student in culinary arts, recently promoted
to Line Cook at nationally known Moosehead Restaurant.

Work Experience

7/08-Present   Line Cook at Moosehead Restaurant, Ithaca, NY
Work at sauté or grill station for lunch or dinner meals
in a well-known restaurant featuring healthful natural foods
cuisine. Perform mise en place and food preparation. Follow safe
and sanitary food procedures. Test and evaluate new recipes. Won
Employee of the Month (June 2009).

10/06-6/08 Preparation Cook at Moosewood Restaurant, Ithaca, NY
Performed all preparation tasks in kitchen emphasizing scratch
cooking and vegetarian dishes. Completed all duties in timely
fashion while maintaining sanitation standards. Received
"Excellent" performance evaluations.

Summers, 2005 and 2006.   Assistant Cook at Lenape Summer Camp,
Seneca Falls, NY. Under Head Cook's direction, did basic food
preparation tasks, cooking, and baking. Assisted in purchasing,
receiving, and inventory management.

Education and Certification

Bachelor of Professional Studies in Culinary Arts Anticipated
May 2009, Olympia University, Ithaca, NY

Dean's List every semester

Treasurer, Culinary Club (sophomore year)

ServSafe® Food Protection Manager, #2364656 (National Restaurant
Association)

Figure 1-8 Scannable Resume

EXERCISES

1. Learn more about resumes at the monster.com website: http://resume .monster.com/resume_samples/

2. Use the Resume Worksheet (Figure 1-4, and on the CD-ROM) to write up the information for your resume. You will probably not include everything you write on this worksheet on the resume itself, so just be complete.

3. After you have gathered the information for your Resume Worksheet, write your first draft using Form 1-5 (on the CD-ROM). Be sure to use action verbs from Table 1-1 and keywords from Table 1-2.

4. Type your rough draft in at least two different formats. The CD-ROM contains five formats. Which looks best?

5. To evaluate your resume, use Table 1-3 or go to the following website and use their checklist: http://www.quintcareers.com/resume_critique_worksheet.html

PROFILE
Lee Cockerell

Learning from those who have blazed the trail before us still proves to be one of the best educational approaches besides formal instruction. Leaders are quoted in trade magazines, speak at numerous affairs, set trends—and stop them—helping establish industry direction. As mentors, they have helped many chefs make good decisions that promote success, they have served as a sounding board for important presentations, and much more.

LEE COCKERELL, Executive Vice President of Operations, Walt Disney World® Resort (recently retired)

Q / What is your present position?

A: I am currently Executive Vice President of Operations for the Walt Disney World® Resort in Orlando, Florida. I am responsible for all of the operations, including the four theme parks, the resort hotels, our shopping and dining complex, and our sports and recreation business. I also have responsibility for the operations that support our operations, including security, transportation, engineering and maintenance, textile service, and so on.

Q / And you currently have how many employees?

A: We currently have 50,000 employees, whom we call Cast Members. Everyone at Disney is a Cast Member, including me. We're the largest single-site employer in the United States, and approximately 36,000 of these 50,000 Cast Members are in operations.

Q / **That's a tremendous number to be in charge of. How did you first decide to go into cooking, or did you decide to go into management?**

A: I was one of those young people who had no idea about what I wanted to do when I got out of high school. I asked a friend of mine what his college major was going to be, and he said hotel and restaurant administration. I said, "Okay, I am going to do that too." I'd never even been in a hotel, as growing up on a farm in Oklahoma left no time or money for vacations. I went off to Oklahoma State in 1962 and spent two years there studying hotel and restaurant administration. After my sophomore year, I went off to Lake Tahoe, Nevada, to work at Harvey's Resort and Casino. The first half of the summer I worked in housekeeping doing turndown service in guest rooms. The second half of the summer, I worked in the kitchen. My title was grease man. My job was to roll this cart around and empty the grease from the griddles before they overflowed.

I kind of got back to Oklahoma late (accidentally on purpose) and missed going back to school for my junior year. I really did not like school; so here I was, and in those days you either were in school or were soon drafted. I promptly joined the Army, went to Fort Polk, Louisiana, for basic training, and became a cook. In cook's school, I met another soldier. He asked me what I was going to do since we were being discharged. I told him I had no idea. He told me that he was going to Washington, DC, to open the Washington Hilton, this new hotel opening in three weeks, and that I should come along. This was February of 1965, and little did I know that this would be the beginning of a 40+ year career in the hospitality business.

I walked into the Washington Hilton on February 26, 1965, and applied for a job. They asked me what I wanted to do, and since I had no idea, I said, "How about a job as a Room Service Waiter?" I had seen this on television and noticed that they made good tips. The lady in the personnel office said that those jobs were all filled but that they needed Banquet Servers. I said fine, and off I went to be a Banquet Waiter. I was fortunate to have one of the Banquet Captains take a liking to me, and he taught me everything. I don't even think I had ever seen a cloth napkin before.

I worked as a Banquet Waiter for a couple of years and then had an opportunity to get into a Food and Beverage Control Clerk position, which led to a management training program. This is when I got my first big break. The job paid so little that I had to get a job at night as a waiter in a French restaurant and also another job on weekends for an outside catering company to be able to survive financially. Those were two great experiences as well.

Q / **Had you graduated already from the university?**

A: No. I didn't graduate from college, and I never went back to finish, so I am probably the worst person in the world to give any advice on the advantages of a college degree. I was lucky, and I would not suggest this strategy to young people today. I do joke sometimes that if I had graduated from college that I would have a really good job today! Despite not graduating, I guess I was lucky that some people noticed my drive, work ethic, and positive attitude. I was Mr. Agreeable.

When they asked me to work on New Year's Eve and then be back on New Year's Day morning at 6 AM, I smiled and said, "No problem." Next thing I know, I am promoted. That has been my secret strategy for all of my career.

On the other hand, my son received his undergraduate degree from Boston University and years later his MBA from the Crummer Business School at Rollins College. I would have had a fit if he tried to do it my way without a degree. I really was the one who inspired my son to go back and get his MBA. These days, business is far more complicated than it was back in my day. When I started, the electronic calculator had not even been invented. Now with computers, everything can be calculated to the tenth of a cent. A person in business today needs to understand all of finance, marketing, cost management and productivity, industrial engineering, and on and on. I definitely would recommend for everyone to start and finish college; and if you can, go ahead and get that advanced degree as well. It will pay off for you.

I worked for Hilton Hotels Corporation for eight years. I held the positions of Banquet Server, Food and Beverage Controller, Assistant Food and Beverage Director, and Food and Beverage Director while working at the Washington Hilton, the old Conrad Hilton in Chicago (now the Chicago Hilton), the Waldorf Astoria, the Tarrytown Hilton, and the Los Angeles Hilton.

In 1973, I had the opportunity to join Marriott Hotels. At that time, Marriott had 32 hotels. I spent 17 years with Marriott and saw them grow to over 800 hotels by the time I left in 1990. Today I think they exceed 2,500 hotels or maybe even more. Marriott was great, and this is where I got the best training on how to manage a business. During the 17 years with Marriott, I held the positions of Director of Restaurants, Director of Food and Beverage, Regional Director of Food and Beverage, Area Vice President of Food and Beverage, Vice President of Food and Beverage Planning, and General Manager. I worked for Marriott in Philadelphia, Chicago, Washington, DC, and Springfield, Massachusetts.

I loved my food and beverage career, but I must admit that my very favorite position was that of General Manager of a hotel. In that position I had responsibility for everything from food and beverage to marketing, sales, rooms, and housekeeping. I loved taking care of our guests and the associates who worked for me. Once you have been in the food and beverage business, you can do anything. The challenges in the food and beverage area are tougher than any other area of the hospitality business. All areas can be challenging, but food and beverage is fast-paced, with hundreds of decisions to be made daily. This really gets into your blood. I love making decisions at a moment's notice, so running a hotel was an exciting experience for me. I held that position for two and a half years.

In 1990, the phone rang one day at my hotel, and it was Disney asking me to interview for a position as the Corporate Director of Food and Beverage and Quality Assurance for the Disneyland Resort in Paris. I went home and asked my wife, Priscilla, what she thought. She immediately said, "Let's go." She said, "Lee, you get to work for Disney, they are going to pay you, and we get to live in Paris. If you don't take this opportunity, you will look back in five years and regret it." I went for the interview. I was offered the job, so I resigned from Marriott the next day and was living in France two months later. It turned out to be the best decision that we ever made. Living in France was exciting and actually exhilarating. I didn't speak French, so I learned quickly what it is like to be illiterate. It was difficult at times; but as they say, "It's the difficult times that make you stronger."

I learned a lot about myself on that job. It was the hardest position that I had ever held. When you hear people talk about working 18-hour days, they are usually stretching the truth a bit; but I really did work 18-hour days, from 4:30 AM, finally getting back into bed at 10:30 PM. I must admit that I would not want to work those kinds of hours my whole life, but in a crunch you have to be able to have the stamina and energy to do what is necessary to get the job done. The clock was ticking toward opening on April 12, 1992. After the opening, I was promoted to Vice President of Operations for the six 1,000-room resort hotels. A year later, in 1993, I was promoted to Senior Vice President of Operations for the resort hotels in Orlando at Walt Disney World. A couple of years later, we merged all operations into one operating group, and I was promoted to Senior Vice President for Parks and Resorts; and in 1997, I was promoted to my current position of Executive Vice President of Operations for all of the operations in Orlando.

I have now been in Orlando for 11 years and with Disney for 14 years. My wife and I have moved 11 times in my career. We have enjoyed every place we have lived, with no exceptions. I tell people all the time that it really does not matter where you live, as you spend 99% of your time at work or home anyway. So if you love what you do and your home life is good, then it does not really matter where you live. I might make one exception to that. My son and daughter-in-law and my three grandchildren, Julian, Margot, and Tristan Lee, live in Orlando and only one mile from me, so that would cause me to not move for sure.

I remember telling my wife when I was with Marriott that we would be moving to Philadelphia and her response was one of alarm when she said, "PHILADELPHIA?!?!" Philadelphia turned out to be one of our favorite places to live, and we made a lot of good friends there who we are still in contact with today, 30 years later. So I can't think of one place that we lived that we did not really enjoy. My wife is a saint. She packed up and moved without hesitation each time I was promoted. Marry a saint if you are going to move a lot.

I tell people all of the time that there are three ways that we learn. First is a formal education, then get as many experiences as you can, and last but certainly not least, travel. Those three things make you a well-rounded person. Those three things are very clear to me and important to me. Meeting people in their country or home and getting to know them on a personal basis makes you more tolerant, and you learn that everyone is trying to achieve the same things for themselves and for their families. This is one thing that I know for sure.

Q / It's interesting that you say that. One of the things that we're trying to establish in this book is a career and education path. This is a very diversified industry, with many career avenues to pursue. If you were going into a culinary program, you would want to get all of your resources together to map out a focused plan. Most aspiring students don't think that far in advance.

A: My advice to aspiring young Chefs would be a little different. I would approach it in a different way. If you are going to start out in culinary, you will need to gain a lot of hands-on culinary experience and technical skill. Technical knowledge is the ticket that opens the first door to entry into a career. You will gain your credibility for having strong technical skills and knowledge. This means you can cook a great meal. At least with technical knowledge and experience, you can cook one good meal.

The second thing to focus on is becoming a great manager. Management is defined as the act of controlling. This simply means that you are well organized, that you have a system in place for planning your day and for following up and for doing the right things in the right order. You know the difference between urgent, vital, important, and limited-value tasks, and you do them in the right order, day in and day out. Being organized is the reason that you can manage a large kitchen and put out hundreds of meals a day and keep your payroll and other costs in line. This ensures that you have the right inventories on hand and that you make a profit. A good system that organizes you ensures that you keep your promises and that your follow-up is excellent and reliable.

The next thing you need to be concerned with is keeping up with technical advances so that it will become quickly apparent to you how to apply technology to your business to help with all kinds of business issues, from marketing to cost controls to improving service to improving the quality of your products. Technology will be the answer to many business solutions in the future, so this is one you must pay attention to.

Last, but not least, you must be a great leader to be successful. When I talk about leadership, I am talking about the ability to lead and inspire others so that you have followers. Without a team of followers, you will be cooking alone. Learn to do four things with your fellow employees. Make them feel special, treat them as individuals, show respect to everyone (keep your biases to yourself), and train and educate your teams and know their jobs so they can get ahead. When you inspire your teams and watch your own behavior, you can be assured that they will look after your business even when you are not there. Listen to your people, involve them in the decisions, ask their opinion, help them, teach them, and recognize them for their good work. Tell them every day how much you appreciate them. This is the way to build their self-esteem and self-confidence. You will have a loyal and dedicated team if you do this. They will be more than interested in their jobs. They will be committed, and committed is far different from interested. Committed teams accomplish extraordinary things.

What happens when you don't understand these four things is this: You are a great Chef, a great technician. You can cook a great meal, but you are not a good manager, so you miss deadlines and your costs are out of control and your inventory system does not work and you are frequently out of ingredients. You don't stay up with technology solutions, so your business is not on the leading edge in the areas of taking care of your guests, employees, and business results. You are not a great leader, so you have turnover and people don't want to work with you because you are abusive and egotistical and your team soon lacks motivation caused by you, their so-called leader in name only, and things go from bad to worse.

It's great that the Chef can cook, but the management part is getting everything organized so you have everything to cook. You and your systems must be well organized if you want to stay in business. Lasting leadership is how you get the whole team working together. You have to get the team inspired, feeling good, and wanting to go out every single night to win. If you are a good manager, a great leader, and you are technically competent and pay attention to technical advances, you will do very well.

Someone once asked me, "How did you get the job you have now?" I told them that I believe the main reason that I have been so successful is that I have a positive attitude. I wake up every morning and go to work and I stay positive. I advise every leader to be careful what you say and do, as they are watching you and judging you. You are the key to your success. Most leaders underestimate the impact that they personally have on the people and the business. I always smiled and said "yes" when I was asked to do something, from working every single holiday to cleaning out the grease traps. I did not always want to do those assignments, but they never knew that, and it was not long before I got promoted. I got picked because I had the technical knowledge, but most of all I got picked because I had a positive attitude and great relationships with others—with my boss, for sure.

So when the job opened for the management training program, I was picked because I had a good relationship with my boss and coworkers. When they asked me to do something, I did it without complaint. That is how the real world works—relationships. When you promote people, you don't have to look in their file. We all know who they are. It is in our head!

If you are going to be a great manager and leader, you need to keep balance between your profession and your personal life. You have to make time to be involved in your community. Chefs are asked all the time to be involved in charity and community events. You need to learn to keep all parts of your responsibilities organized. There is enough stress already in the food and beverage business without being disorganized as well.

These young adults that are making their choices now in school need a foundation, and it's not so easy to do hands-on training as it used to be. I learned management by working under good managers, and I learned skills on computers by someone teaching me. It's more complicated today.

Today, just like back then, you have to count on yourself. You were lucky years ago to have good managers and leaders to learn from. Everyone is not going to be lucky, so they have to count on themselves, as they may not end up with a good mentor to guide them along. I, too, am grateful for the good leaders that I had along the way, and I can tell you I had some pretty bad ones too who I learned a lot from also. My advice is for people to make sure they have the technical skills mastered by going to the right programs or schools and by getting the right experience. Take classes and read a lot about management and leadership and focus on learning about technology. If you have a great leader to learn from, then you are lucky; but most people can make their own luck with hard work, planning, and thinking.

Q / What is a typical day for you?

A: My days are very routine. I get up at 5 AM. I make my wife's coffee so it is ready when she wakes up. I go to Einstein's and have my breakfast, which now is a low-fat yogurt and a cup of coffee with cream and sugar. I read USA Today, and I get to my office by 6 AM. For the next couple of hours, until 8 AM, I write my weekly newspaper, The Main Street Diary, *which goes out to all 50,000 Cast Members on Friday night at 5 PM. I do my email, I plan my day, and I clean up my mail from the day before. My first appointment is at 8 or 8:30. I am driven by my schedule. If something is important, I schedule it, including walking the parks and resorts and other operations. I schedule time to talk with guests and with Cast Members and to do things like teach a time management course once a month. The minute something is important to me, it gets scheduled. Someone taught me long ago to schedule the priorities in your life. That is why my workouts are scheduled appointments in my calendar. I have a full day of meetings, appointments, and walks of our property. I stop working at 5 PM and go work out and stretch at one of our spas until 6:45 and get home around 7 PM. My wife and I have dinner, and we hit the sack pretty early Sunday through Thursday.*

On Fridays we go to our beach house, which is 90 minutes from Orlando on the Atlantic, and spend the weekend with my son, Daniel, his wife, Valerie, and our three grandchildren. Some people may think that routine is boring, but I have learned that routine is really important. If you want to know what is going on, maintain consistency in your business. We also take our grandchildren to Disney frequently and visit just like any other guest. We wait in line and we do the parks just like a visitor, for me to learn the truth and for the kids to have fun. If you have routine in your life and schedule your priorities, you will find yourself doing what you are supposed to do versus just what you like to do.

Q / How do you communicate with your 50,000 Cast Members and get your philosophy to trickle down?

A: Actually, I worry about my philosophy trickling up and down and all around. I have several strategies on how I do this. When I came to Walt Disney World, in 1993, I worried about this one thing a lot. How was I going to get all of these people to understand what we needed to do? First, I personally write a weekly newspaper for all 50,000 Cast Members. It goes out every Friday at 5 PM. In that paper are numerous columns that I write on leadership expectations, our purpose and role, our vision and how we achieve it, diversity, pre-shift meetings with our teams, general important information, and on and on. This paper is used to not only inform but also to recognize our Cast for the great work they do. We print numerous guest letters that compliment the Cast on the great job they do.

I hold monthly meetings with our front-line Cast and our front-line management to listen to them and to explain things to them that are concerns.

I make myself available to speak to any group so I have the opportunity to deliver my messages in person. I speak to thousands of people a year.

I make myself available to see anyone who wants to see me, as do all of our executives.

To deliver my messages, I teach classes on leadership, time management, and how to build commitment on a monthly basis.

I visit the operations frequently and talk to the Cast about what is important. I inspect the restrooms for cleanliness and check the food. I check the break rooms and cafeterias. I make unannounced visits as well.

I pretty much know what is going on, as I talk a lot with front-line Cast who have direct interaction with our guests. After a while, you get a great reputation for this sort of thing, and then the Cast tells you everything. They actually help me do my job better than anyone else because they know why we are trying to do what we do and how we do it.

To really know what is going on and to get your messages and expectations embedded, you have to have a clear and routine strategy in place. You need a few simple messages that you deliver over and over and over until you are blue in the face and then deliver them again.

Our Cast Members read. They give copies to their friends and neighbors and their children. We even have articles in there on how to raise your children and how to deal with things like getting homework done and how to properly discipline your children, and how to build your relationship with your partner or spouse. We talk about our expectation for being fair and firm. We talk about our expectation to treat everyone with respect no matter what their background, color, race, culture, or sexual orientation. Respect, appreciate, and value everyone is the battle cry around here for how to treat people.

Q/ So I guess that they know you are going to take action and do something about it. Is that correct?

A: That's right. They call me too. I have a confidential voice mail number that I constantly publish. Anyone can leave me a message and whether they leave their name or not, that is up to them. I follow up on every single item. A manager in my office traces every single item until it is resolved one way or the other. This is why I have such credibility. I follow up! When they don't leave their name, I put their message in the Main Street Diary. *I say, "This week*

I got a message about the locker rooms not being clean. Since you did not leave your name, I just wanted you to know that I took care of that and I hope that you are happy. Call me back if they don't stay clean." We just keep pounding away on stuff like this. These are the kinds of things that are important to people. The leaders behave better too, because I tell them if your leader is not behaving, let me know.

Q/ How do you motivate and educate yourself? You are the leader, the main visionary.

A: When I was young, I was very introverted. Actually, I took a speech course in college and I dropped the course the night before I had to give the speech because I was so terrified. I was a quiet, shy little boy. Seventeen years later, I am in an executive position, and I am asked to give a speech to 300 guests of the Chicago Marriott, where I was working as the Director of Food and Beverage. I agreed. What I forgot is that I did not know how to give a speech. I wrote out something on a yellow pad and went out there and made a fool of myself. That fear came rushing back. After that day, I went to get some help from an expert who happened to be Bill Marriott's father-in-law, who taught speech in college. He gave me some of the best advice. He told me, "Lee, always talk about things you are passionate about; use personal examples; and always prepare your own material. Don't let people write speeches for you. Talk about your kids, your mother, your dog, or whatever." So I started doing that, and it works beautifully. What he was telling me was to tell stories and not make speeches. People don't remember speeches but they do remember stories and the people who tell them, therefore they remember the lesson that is taught by the story.

Q/ Right. You have such a reputation here for speaking that anybody who comes in contact with you usually has a notepad with them and they're taking notes.

A: I tell leaders to learn how to be good communicators by watching people who are good at giving speeches in addition to taking classes, and take the time to test out your speaking skills on your staff. This is what I did. Create a few message points that you want to become known for and talk about them all the time. You can tell lots of different stories to make the same points. One day it occurred to me that leaders basically talk for a living. We try to figure out what is going on, and then we figure out what we want to be going on, and then we communicate with our teams to try to get them to do what we want them to do. We communicate in writing, and we communicate by speaking. I think speaking is the best method, as there are far fewer misunderstandings when you speak to people because they can ask questions to clarify.

Experience is another way to develop yourself. For example, I had a mentor once who was the Director of Food and Beverage of the Waldorf Astoria. His name was Eugene Scanlan. He later became the General Manager of the Waldorf Astoria. He was the first Executive Chef there after starting as a young apprentice at 17. He was very impressive. He would take me and another young manager for dinner every Monday night to one of the restaurants in the hotel, and he would make sure that we ordered different things, or he would order them for us. He would have us taste each dish, and then he would explain the dish including the ingredients, how it was prepared, and any other history of that dish. He would order different wines and do the same. The first time he ordered raw oysters, I was wishing I was not there. I am from Oklahoma, and I had never eaten a raw oyster. When I looked at it, I was pretty sure I would never like it. I really didn't want to try it; but with Gene's insistence, I ate it, and I liked it. From those early experiences, I learned a lot. First, I learned that it is important to mentor others. This was a real gift that was priceless. To this day, I still encourage others to try things and to get varied experiences. I tell people that there is more to food and beverage than a cheeseburger and a beer, and that there are a lot of other wines besides Cabernet Sauvignon and Chardonnay. This, like all experiences, gets better and better with time. Just think about how much you would miss out on if you did not try things. Just like those oysters that I thought I did not like, it is the same with public speaking. When I first started doing it, I hated it. Now I love it and want to do it as much as possible. In both cases, I had great teachers who showed me the way.

Q/ Who do you report to?

A: I report to the President of Walt Disney World. He has been here since he was 17. He started in a front-line position in Cash Control ringing out the registers at the end of the night. In those days, it was called a Z run. Today, he is the President. It pays early on at least to get with a company that is well-known, I think. I would tell any graduate to go to work for a large well-known quality company when they get out of school and stay five years. In those five years, you will get great training and experience. If at the end of those five years you are not achieving your goals, then move on. A well-known company will open many doors for you for your next move. Don't go jumping around every year from company to company for a few more dollars. You want stability on your resume. After five years, you will still have at least 40 years to work. Get really good at what you do before you start moving around.

Here is some excellent advice that my boss gave to someone recently. One of our executives went in and said, "Now that I've become a General Manager, how do I get ready for my next position and promotion? What is your advice?" He said, "I wouldn't worry too much about what you have to do to get the next position. What you should be worrying about is what you have to do now to get that really big job ten years from now." You might have to go back and get your MBA or get certain experiences under your vest over the next ten years. This is a great question, I think. What do you need to do today, this week, or this month that will pay off for 5 to 40 years from now? One thing comes to mind for me and that is to pay attention to your health. And a really big and important one is that if you smoke . . . stop!

Q/ Were you always an excellent leader?

A: No, I was not. When I first started my career, I was a great manager because I was so organized that there was no deadline that I could not make. I was a highly disciplined person with a system for keeping on top of things. I was very organized. The problem was that I was so focused on getting things done that I did not focus or pay attention to people. I forgot that it all gets done through people. I know that today, and I use my organizational skills to pay attention to the work and to the people. I am now a great manager and a great leader.

Being aggressive and not being focused on people can get you into a lot of trouble even though you might get short-term results with that style. I got fired once after 90 days on the job. I should never have taken the job in the first place, as it turned out that the place was going bankrupt, which they forgot to tell me in the interview. My wife had told me not to take the job in the first place, but of course, I was pretty aggressive and knew everything, so I took it. This was a good lesson in listening to others. I went home and told my wife, Priscilla, that I got fired. She said, "Good, I hate living here." Later on in my career, I got passed over for a big promotion, and it turned out to be because of my aggressive style which was to get things done at the expense of people sometimes. I was more focused on getting things done than I was on developing and inspiring people to get things done. This was in 1985. I really knew then that I had to take a long, hard look at my management and leadership style. I started studying leadership. I went to seminars. I read a lot about leadership. I thought a lot about my behaviors and how they affected people. I became totally aware of myself and the impact that I had on people. I made a lot of changes. I learned how to listen, how to show respect to others, how to involve others in the decisions, how to build others' self-confidence and self-esteem. I am a lot better today than I was then, and I continue to get better. I even became a better leader for my family and friends as well.

I can guarantee you that in the first half of my career, there were thousands of people who couldn't stand me, and now in the second half of my career, there are thousands who have a great deal of respect for my leadership. I just wish I could get them all together for a weekend to clean up my reputation from the early days! Today, I have a great amount of respect and admiration for the people I lead, and I make sure that I let them know that. In the old days, I thought that people had to listen to me. I have learned that it is the other way around if I want to get great results. I now know that I serve them and not that they serve me.

Being in a place where you are learning is really important too. I go out to the operations a lot. One day I asked one of our cooks how long he had been in one of our restaurants, and he told me he had been there since it opened. I asked him why he was still there, and he told me that he stayed because the Chef taught him something new every day. That is a great lesson for all leaders. Develop your people. Get them ready so they can move up to the next level and have a better life.

I am doing a lot of research right now on the subject of commitment. How do leaders inspire commitment from their teams? I know for sure that the following things are important: (1) Make your employees feel special.

(2) Treat your employees as individuals. (3) Treat everyone with respect. (4) Develop your employees, teach them, and know their jobs. Do these things, and you will have a committed team who will do anything for you, and they will go all the way. Being interested in your job and being committed are two different things.

I think a lot about my responsibility to create an environment where people are happy. Happy people don't quit, and they live longer. My main responsibility at Walt Disney World is to create a happy, healthy environment where every single Cast Member can achieve whatever they are capable of. Get people into jobs that they can be happy in. Many unhappy people are just in the wrong positions.

Q/ What do you see as trends in the industry?

A: The industry will always be changing. I think we've got to have an industry where people can have a balanced life. The days of working 14 and 15 hours a day and six and seven days each week are over. I make sure that our people get two days off a week, and we expect a ten-hour workday on average. That's enough. Go home and have a life. We really insist that our people get off to do what they have to do. Go see your son or daughter in a school program or to that teacher's meeting or whatever. Don't miss these things and then have regrets some day. This is one reason that we have such low turnover.

Q/ In your organization there must be so many job opportunities: teachers, Chefs, Sous Chefs, and so on.

A: Yes, there are, and add to that purchasing, test kitchens, restaurant managers, and on and on. We have 7,000 leadership positions at Walt Disney World. There are lots of opportunities here.

Q/ In the test kitchens, would Chefs develop new menus? Is that a whole division?

A: We have our Chefs work together to create new things. We have a team just working on coffee ideas. We have a team just working on desserts and others that just think about children's menus, and others on healthy eating, which is a big craze again right now. We have people just focused on wines. We make over 1,500 wedding cakes a year. You need a lot of culinary and artistic talent to get all of this done.

Q/ What is the volume of food and beverage?

A: This year, we did just about $1 billion in food and beverage sales at Walt Disney World® Resort in Orlando, with over 100,000,000 food and beverage transactions from full service to quick service to buffets to carts and snack bars. We have hot dog carts that do $5,000 a day. I wish I owned that cart. Walt Disney World is an exciting and magical place to work, and I just love working here. This is a place where we really do make dreams come true and a place where everyone who works and plays here finds a real sense of joy and inspiration.

CHAPTER 2

Put Together a Job Search Portfolio

Introduction

FOR YEARS, CERTAIN PROFESSIONALS, such as artists and writers, have used portfolios in their job searches to showcase their abilities and qualities. If you have ever seen an artist's portfolio, you know that it contains an organized collection of his or her work based on artistic style, growth, abilities, and aspirations. By the time you finish looking at the collection, you have an excellent idea of who the artist is and what he or she does well.

Portfolios aren't just for artists and writers. Professional portfolios can be used by anyone who has a career to:

- Get a new job
- Get a raise
- Get a promotion
- Help direct and keep records of lifelong learning

In addition, you may be asked to maintain a learning portfolio in college. A learning portfolio, which tends to be quite long, documents learning in specific areas, such as sanitation. A job search portfolio focuses on your work-related skills, abilities, and qualities that are necessary to do a job. This type of portfolio is much shorter, from 10 to 30 pages in length.

Think of a job search portfolio as a catalog of your skills, abilities, and qualities. It is not simply a resume. A portfolio goes beyond your resume to demonstrate what is on your resume and be a visual representation of your strengths. Artifacts such as menus and photographs of plated meals are a visual way to demonstrate your accomplishments. Just as archaeologists use artifacts to reconstruct a civilization, a portfolio reconstructs your career.

Why bother with a job search portfolio? Assembling a portfolio has many benefits:

◆ It helps you assess your learning and work experience and compare it to an employer's need for skilled, capable employees.
◆ It helps you prepare for interviews.
◆ It gives you a competitive edge when applying for a job because it showcases and convinces others of your skills, abilities, and qualities, and it demonstrates the results of your work.
◆ It helps you communicate clearly.
◆ It provides evidence of your potential.

Creating your own job search portfolio is a fairly simple process, as we now show.

Do I Really Need a Portfolio?

That's a very good question that has no right or wrong answer. You need to gather the facts and make your own decision, keeping in mind when interviewing the timing to present a portfolio needs to be right. You can approach this by asking if the interviewer would like to see samples of some of the information you are providing. Keep it short and sweet with your best work, presented in a clean, organized professional manner.

When I think of this word, PORTFOLIO, I envision a large black leather folder that an artist takes on interviews or to a curator. The fact of the matter is, we as culinarians are artists in our own right. There is a tremendous amount of our work that just, can not be explained by words alone. A physical menu can better represent your cooking philosophy, better then trying to explain your style while getting tongue tied because you ran out of adjectives. Talking about yourself in an interview with some well presented, hand-selected representation of your work and style is a powerful tool that can act as a support of confidence during the grueling act of interviewing.

Some examples of this are:

◆ Samplings of daily specials that you either prepared or designed
◆ Special event menus
◆ Quality photography of plates, buffets or centerpieces. This may be an opportunity to get a professional photographer who can capture these special event
◆ Menus that you were a part of both in the creative design and the cooking execution. Be careful not to include information that you are not knowledgeable about, and say things like, "I was only the Sous Chef, the chef did all the menu writing. I cooked some of the stuff." You want to be well versed in all material you include; remember this is your portfolio.

- Press which you have gotten, newspaper articles, magazines, etc. This is a great symbol of your accomplishments, because those editorials are hard to falsify.
- Copies of awards, certificates, certifications
- Updated reference list — and I mean updated

In reality not many people will ask for this type of information, but it is a great exercise for us as professional to keep our accomplishments in a book to reference or to document your activities. Another smart way to approach a portfolio, giving the times being the way they are, is to create a power point presentation with all your information. It may be much more appreciated by a forward thinking chef. Whatever your decision, it is still smart to create a place to store these documents, pictures and/or articles for the times when you will need to locate them.

Choose How to Organize

You can organize your job search portfolio in a number of ways. Here are some possibilities:

- **By resume category** — Probably the most popular and easy-to-use method, the resume-based portfolio provides artifacts for each of your resume categories. See Table 2–1 for a list of potential resume categories.
- **By date (chronological)** — With a chronological format, a portfolio is normally divided into the years you were in college, the years you worked for ABC Restaurant, the years you worked for XYZ Catering, and so on. This portfolio is essentially divided up by jobs you have had.
- **By skill (functional)** — A functional portfolio is organized according to your skills. For example, an experienced cook may include the following skills categories: cooking, garnishing, catering, sanitation, and technology.
- **By theme (thematic)** — A thematic portfolio is divided just the way this sounds—by theme. For example, a chef working in a catering company may organize the portfolio into sections based on types of catering affairs: weddings, corporate affairs, and so on.

Table 2-1 Potential Categories to Use in a Resume-Based Portfolio		
Professional Experience	Computer Skills	Publications
Education	Foreign Language Skills	Presentations
Certifications	Community	Military Service
Continuing Education	Service/Volunteer Work	
Professional Affiliations	Awards/Honors	

So which one do you pick? Whichever format will encompass the artifacts you need to exhibit to get the job you want. Think about the requirements for the job you are looking for, and then choose a format that helps prove you are the best match for the job.

Whichever format you choose, you should include a resume and reference list at the front of your portfolio. You may also want to include a work philosophy or philosophy of cooking statement. This is a description of the guiding principles that drive you and your cooking, including your philosophy of foods and cooking, your work ethic, management philosophy, and so on. Your cooking philosophy may be, in brief, to emphasize local, organic foods in simple meals, or to blend traditional with contemporary cooking. In a healthcare setting, your cooking philosophy may be to provide home-style, attractive meals that patients enjoy. In any case, your work philosophy should be about one paragraph long. It could even be put into a bulleted list of three to five points.

Collect the Contents

Next, you need to collect and select artifacts for the portfolio. To do this:

- Think about what you do, the skills involved, as well as how you do it. For example, if one of your job duties is to cook for banquets, you could select a banquet menu and photographs to show your cooking and food presentation skills.
- Collect artifacts that match the requirements of the job you are looking for.
- Choose items that are your best examples and show mastery. You don't want to include average or mediocre results.
- Try to find examples of your work in which you were the only or at least the major contributor.
- If you are still in college, or only out a short time, don't forget that you have developed work-related skills while playing team sports, performing volunteer work, engaging in hobbies, and going to school. For example, being a member of a sports team requires discipline, motivation, teamwork, and energy. Performing volunteer work shows dedication and can develop any number of skills. Engaging in hobbies shows motivation and skills. Working on team projects at school requires teamwork, problem-solving skills, and people skills. It's fine to include these activities and skills in your portfolio.

Here are examples of elements you might select for your portfolio. They are divided into resume categories.

- **Professional Experience**—This section is likely the heart of your portfolio. Table 2–2 shows categories of skills and gives you more ideas for artifacts. You may want to subdivide Professional Experience into skill groups if you have more work experience. Table 2–3 gives examples of personal qualities you may want to highlight in your portfolio.

Table 2-2 Possible Skill Categories

Communication Skills	Management Skills
Creative Skills	Marketing and Sales Skills
Culinary Skills	Supervisory Skills
Financial Skills	Technology Skills
Human Resource Management Skills	Sanitation and HACCP Skills

Table 2-3 Personal Qualities to Highlight in a Portfolio

Accurate	Flexible	Punctual
Adaptable	Hardworking	Reflective
Careful	Honest	Reliable
Cheerful	Imaginative	Resourceful
Confident	Innovative	Self-starter
Cooperative	Logical	Sensible
Courteous	Loyal	Sensitive
Creative	Motivated	Steady
Efficient	Open-minded	Tactful
Energetic	Patient	Trustworthy
Enthusiastic	Persistent	
Ethical	Practical	

- Photographs of plated foods, buffet tables, etc.
- Samples of menus
- Samples of recipes
- Customer survey results
- Performance evaluations
- Documentation of accomplishments: increases in sales, decrease in costs, etc. (Can use bar graphs, pie charts, or other graphics)
- Major projects completed
- Training materials developed
- Title page of report written
- Newspaper/magazine clipping describing event you contributed to
- Thank-you letters from customers
- Menu clip-ons or other artifacts of marketing/sales skills
- Memos, reports, letters that show communication skills
- Education
  - Copy of your college diploma(s)
  - College transcript

- College course descriptions
- Copy of scholarship letters
- Copy of awards or honors
- Copy of honor society memberships
- Photographs or other artifacts/internships
- Photographs or other artifacts/extracurricular activities
- Listing of leaders of an organization in which you held an office
- Photograph or other artifacts of service project participation
- Title page or other artifacts of relevant course projects such as a business plan
- Certifications
  - Copy of certificates
  - Description of what you are doing to recertify
- Continuing Education
  - Chronological list of workshops, seminars, and courses attended during last five years
  - Certificates (if available) from workshops, seminars, and courses
  - Chronological list of trade and industry shows attended in last five years
- Professional Affiliations
  - List of organizations to which you belong, the year you joined, offices you held, and boards or committees on which you served
  - Copies of current membership cards
  - Artifacts of leadership positions held
- Computer Skills
  - Copies of menus, brochures, etc., you created
  - List of computer software you are proficient in
- Foreign Language Skills
  - List of college courses you have taken in a foreign language
  - Statement of which languages you are proficient in, and whether you are proficient at reading, writing, or speaking each one
- Community Service/Volunteer Work
  - Photographs, flyers, menus, or other artifacts of volunteer work
  - Awards/Honors
  - Copy of certificate or photo of you accepting award/honor
- Publications
  - Title page of published articles
  - List of publications, including date published and publisher's name
  - Front cover of publications
- Presentations
  - List of presentations, including date, topic, and place
  - Sample of visual aids, handouts, etc.
- Military Service
  - Copy of honorable discharge showing years of service
  - Verification of military education
  - Certificates verifying competence in various areas

- ◆ Commendations and awards
- ◆ Photographs of accomplishments

When collecting artifacts from work, include only items you clearly own or have permission to include. You do not want to reveal proprietary information about your current or past employers.

If you want to develop a functional, or skills-based, portfolio, use Table 2–2 as a starting point to develop your own skill categories.

Get Supplies

To put together a job search portfolio, you will need supplies.

1. **A slim three-ring loose-leaf binder with inside pockets:** A zippered binder is a good idea if anything could accidentally fall out of the binder. A view-binder, which has a place to insert a front cover you create, may also be a good idea.

2. **Sheet protectors or insertable plastic pockets:** Sheet protectors are great for storing many of your artifacts. They also make it easy to add and remove pages from the binder.

3. **Tab dividers:** Make a tab to identify each category of your binder. Because sheet protectors are larger than 8½-x-11-inch paper, buy extra-wide tab dividers so the tabs can be easily seen and read.

4. **Plastic photo sheet holders:** Plastic sheets that hold several photographs are helpful when you want to display more than one or two photographs on a page.

You will also need heavy (24#) bright white paper to mount many of your artifacts and card stock for your titles and captions.

Put the Portfolio Together

Before you assemble your portfolio, remember that more is not better. A job search portfolio is a professional portfolio, not an overstuffed photo album. The following steps will guide you through the process of finishing your portfolio. You will also find them on the CD-ROM as the Portfolio Worksheet.

1. Decide which type of portfolio, or which mix of portfolio types, will best serve your purposes. There is no one correct way to organize a portfolio, so choose what works for you.

2. Come up with a tentative list of categories to include in your portfolio.

3. From the artifacts you have collected, pick out the best sample(s) for each category. Keep in mind that it's better to have a few good samples than lots of mediocre or redundant ones, and that an interviewer can absorb only six to eight samples. If you see you have too much for one category, you can either split it up or be pickier about what you select. If you have too little, don't worry. One or two artifacts are probably enough in areas not directly related to work. You may also be able to combine two categories, such as certifications and professional affiliations.

4. Decide if you want your first tab to be for your resume and reference list. If you prefer, you can put these documents in the pocket of the loose-leaf binder. Decide if you want to include a work philosophy in your portfolio. If so, put it at the front.

5. Create an index tab for each category.

6. Prepare the artifacts to go into the binder. For each item, such as a photograph or a menu you created, you need to write a title and caption. After you think of a concise title for each artifact, work on the caption. Your caption can be in the form of a short paragraph or a bulleted list. When writing your caption, consider these questions:

    ◆ What is being shown? What were the results?
    ◆ How did you accomplish this task? What skills did you use? What personal qualities, such as persistence and flexibility, were important?
    ◆ Why were you doing this? Why is it important?
    ◆ When did this happen?
    ◆ Who worked with you on this?
    ◆ If anything presented is confidential, do you have permission to use it?

   See Figures 2-1 and 2-2 for examples of titles and captions.

7. Now you can start laying out your pages. A portfolio contains diverse artifacts and will not look professional unless you unify its presentation by being consistent in your layout. As with your resume, allow 1-inch margins on the sides, top, and bottom of each page. Pick a simple typeface that is easy to read. Don't clutter your pages; lots of white space makes them easier to read. White space is the space on a page that is not occupied by text or pictures. Avoid headlines in all capital letters. It's actually easier to read headlines in lowercase with the first letter of each word capitalized.

8. If your artifact consists of one or two photos, you can tape them on the paper after you have printed the headline and caption. Place the headline at the top of the paper and the captions along the sides of the pictures or at the bottom (see Figures 2-1 and 2-2). Whichever style you choose, be consistent. If your artifact is a menu you want to put in a sheet protector, print your title and caption on card stock, cut it out, and let it float on top of the menu in the sheet protector. If you use photo sheet holders, use card stock to identify the photos and put it with the photo if room permits. You can

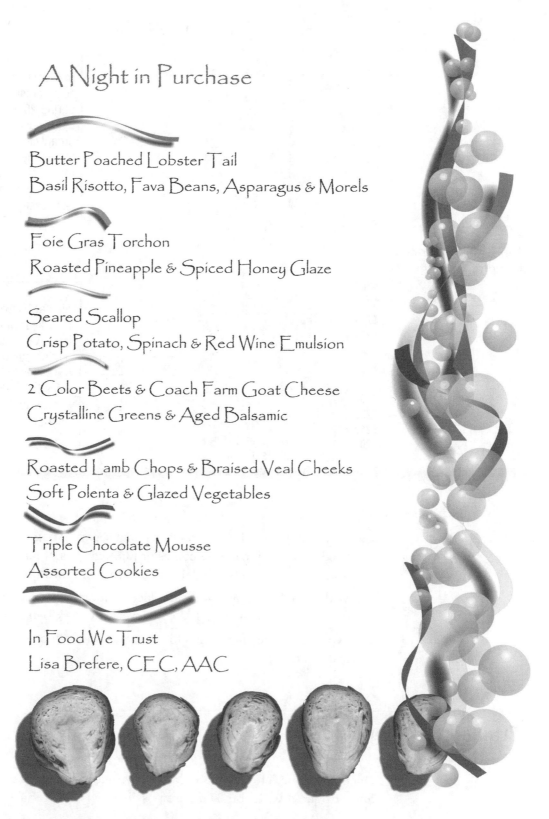

A Night in Purchase

Butter Poached Lobster Tail
Basil Risotto, Fava Beans, Asparagus & Morels

Foie Gras Torchon
Roasted Pineapple & Spiced Honey Glaze

Seared Scallop
Crisp Potato, Spinach & Red Wine Emulsion

2 Color Beets & Coach Farm Goat Cheese
Crystalline Greens & Aged Balsamic

Roasted Lamb Chops & Braised Veal Cheeks
Soft Polenta & Glazed Vegetables

Triple Chocolate Mousse
Assorted Cookies

In Food We Trust
Lisa Brefere, CEC, AAC

Figure 2-1 Sample Portfolio Page with One Caption

Financial Results

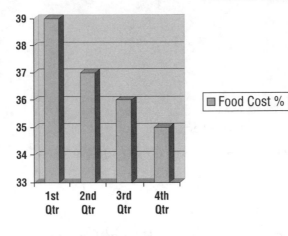

While I was the Chef at Midwest Medical Center, I lowered food cost percent from 39 to 35 during 2008. To accomplish this, I joined a group purchasing organization and used bids more frequently. By better training the receivers, we tightened up our receiving procedures and had less waste.

Figure 2-2

Sample Portfolio Page with Two Captions

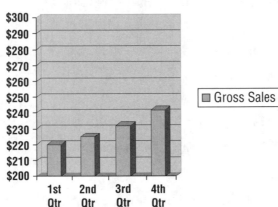

Also at Midwest Medical Center, I worked with the cooks to improve the Café menu. By also getting customer feedback, we were able to increase customer sales from $220,000 in the first quarter to $242,000 in the fourth quarter.

always put the titles and captions on a separate page if needed. You may need to use reduced-size copies or photographs of large and bulky three-dimensional objects (medals, etc.) to fit them into the binder.

9. Within each category, you can place your artifacts in chronological order or put your best ones first. Do whichever will work better in an interview.

10. Type up a table of contents for your portfolio. Place it up front in the loose-leaf binder.

Evaluate Your Portfolio

Once you complete your job search portfolio, ask a friend or teacher to review and critique it with you. Here are possible review criteria:

◆ Is the job search portfolio neat and organized?

◆ Is the portfolio labeled properly?

- Is it easy to use?
- Does your portfolio market your talents?
- Does your portfolio highlight what you need to secure a job?
- Are the titles concise?
- Are the captions engaging and polished?
- Do your artifacts support your knowledge, skills, abilities, and qualities?
- Have you avoided using jargon, acronyms, and abbreviations people won't understand?
- Have you proofread it for spelling and grammatical errors?

Use a Portfolio in an Interview

Before you go into an interview, make sure you have the statement "Portfolio Available" at the bottom of your resume. With luck, the interviewer has read that and remembers the information.

During the interview, look for appropriate opportunities to show parts of your portfolio. For instance, when you are asked about your experiences making food attractive, tell the interviewer that you have a portfolio. Turn to the appropriate section in your portfolio and show the interviewer a page or two. If you are seated at a table with the interviewer, you can place your portfolio in front of him or her. Otherwise, you may want to remove the page from the binder and hand it to the interviewer. Artifacts have only a few lines of text at most, so you need to chat about them, especially in relation to the interview question. Here are additional guidelines:

- Don't narrate your portfolio page by page during the interview. Instead, draw on it only as the interviewer's questions come up. If the interviewer asks for an overview of your portfolio, briefly describe how it expresses your work philosophy and how it is organized. This will enable the interviewer to know enough to question you further.
- You can change the contents of your portfolio based on the job for which you are applying. For example, if you are a Chef in a restaurant and you are looking for a Catering Chef position, you may want to add more quality artifacts to the Catering Skills category.
- Watch to see how interested the interviewer is in your portfolio. Try to gauge the level of interest. If the interviewer does not show a lot of interest, don't draw on the portfolio more than once or twice.
- Know that you will have interviews in which you do not use your portfolio at all. Some interviews are mainly to determine if you fit in well with the other players. A portfolio is more useful when the interviewer is still trying to ascertain if you have the knowledge, skills, abilities, and qualities to do the job.

Keep Your Portfolio Up to Date

Once you have gone to the trouble of assembling a job search portfolio, spend the time to keep it up to date. Review your portfolio two or three times a year. Remove anything that is out of date, and add new artifacts (with titles and captions) to the appropriate categories. Your objective is to keep your portfolio as responsive to future needs as possible.

EXERCISES

1. Go to the following website: http://www.acinet.org/acinet/crl/library.aspx. Type "portfolio" into the keyword search and select one of the many excellent resources on portfolios. Read one of the resources and write a paragraph on what you learned.

2. Collect artifacts for a job search portfolio. Next, follow the steps to putting together your portfolio, using the CD-ROM.

3. Have a friend or teacher review and critique your portfolio. Make changes as needed.

4. Evaluate your portfolio using the Evaluation of Portfolio found on the CD-ROM.

5. In a mock interview, use your portfolio to support your candidacy.

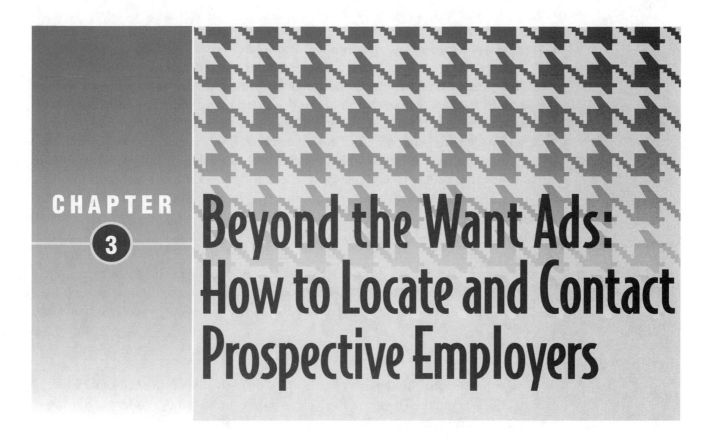

CHAPTER
3

Beyond the Want Ads: How to Locate and Contact Prospective Employers

Introduction

ALTHOUGH THE JOB ADVERTISEMENTS in the Sunday newspaper seem to offer lots of positions, trying to find one only by wading through the want ads does not often work—especially as your salary requirements rise. This is because most jobs are never advertised. Research indicates that one of the most effective ways of finding out about jobs is by getting leads from people you know, a technique called networking. Many more positions are filled every day through word of mouth than through printed advertisements. Another effective way to find a job is to identify potential employers yourself and call them or knock on their door. This chapter discusses a variety of ways to find jobs. As you read, keep in mind that the more methods you use to find a job, the greater your odds of getting a job offer.

General Guidelines

You may find it frightening to look for a job because it means calling strangers, being interviewed, taking risks. While the experience may pull you out of your comfort zone, it is clearly an opportunity to meet new people, grow, and become more confident. The worst that can happen is that someone may not want to help you in your job search, may not want to interview you, or may give the job to someone else. The best that can happen is that you will find a job you love.

Looking for a job requires time and organization. Be prepared to spend at least 20 hours each week using the methods discussed in this chapter to find out where job openings exist, get your foot in the door, and be interviewed. You will be busy emailing, faxing, snail mailing, and delivering your resume to lots of potential employers (many of whom will never call you). Set aside certain times every day to follow up leads, make phone calls, check for job openings, and so on.

To be successful with your job hunt, you must be organized. One way to organize is to write up an index card for each job you applied for. Write notes on everything about the job and the company on the card. Keep the index cards in alphabetical order by the phone. If you get a phone call from one of the companies, all your information is right there. You can also keep track of resumes you send out, phone calls, and other communications by maintaining a job log. The CD-ROM contains a job log that you can use to review progress in your job search.

Here are more suggestions:

◆ Use a variety of strategies to find a job. Don't rely just on the Internet or just on the want ads. The more strategies you use and the more time you put into your search, the more likely you are to find a job.

◆ Don't get discouraged if you don't get a job within a month. Short job searches can take up to four months.

◆ Learn about the companies you have applied to or want to work for.

◆ Make an effort to contact and establish a relationship with the hiring manager. Although you may think you are being a nuisance, you're not. Your initiative and persistence are positive traits.

◆ Always follow up after you have sent someone your resume. If you email your resume to someone, mail a paper copy with a cover letter to the person that same day. Then, within a week, if you have not been contacted, call the person to check if your resume arrived and if you can set up an interview. Reiterate your strong interest in the job and the company.

◆ Be persistent. If the restaurant said last month that there were no openings, go back and find out if the situation has changed. If Human Resources said you should call back in a week, make the call.

◆ Think beyond the employers you are currently considering to additional employment possibilities. Sometimes jobseekers are convinced they will get a job from the handful of employers they are seriously looking at. Don't put your job search on hold while you wait. You are wasting valuable time that could be spent looking for other jobs.

◆ Don't limit your search to a certain dining segment such as fine dining. Any type of foodservice is a great place to develop preparation skills. Mastering the quick pace of grill, sauté, and pantry can be of value in many jobs with fabulous menus and talented Chefs. A banquet cook working in an upscale catering house can apply the precious skills he or she is learning in many other settings.

- ◆ The more people you talk to on the phone and visit in person each day, the quicker you will find a job.
- ◆ For smaller restaurants, it's much better to go in person rather than use the telephone. If the restaurant or company is big enough to have a Human Resources department, then it is appropriate to call. In either case, avoid calling at times when the staff are likely to be busy (mealtimes) or otherwise unavailable.
- ◆ Be prepared to tell employers why you are the best choice for the job and what makes you different from other candidates.
- ◆ Always write thank-you notes after interviews (see Chapter 5).
- ◆ Always treat everyone courteously during your job search. Even if no jobs are available, someone may still be able to refer you elsewhere, so you always need to make a good impression.

The final guideline is to accept rejection. It's a normal part of the job search process. Don't take it personally. It doesn't mean you won't find a job. You will; it is just going to take more work.

Search the Hidden Job Market

The hidden job market includes jobs that are not advertised in the Sunday paper, posted on the Internet, or otherwise made public. Networking is by far the best way to find jobs on the hidden job market. Another way to uncover job openings is to target and contact specific employers.

NETWORKING

Networking is not only helpful in finding jobs but crucial for advancing in your career. What is networking? Networking is investing regular time and effort to create and maintain career-related contacts. It is building relationships and sharing information. What does networking look like?

- ◆ Networking involves talking with people at work and outside of work to gather information, ask for advice, or learn about something new in the culinary field.
- ◆ Besides networking face-to-face, you can also network on the phone, via email and electronic bulletin boards, and by letter; however, talking in person is most effective.
- ◆ Networking is a two-way street. You get useful information and contacts, and in turn, you act as a resource and contact. Networking involves give and take: Expect to give as much as you get.

- You don't need special skills or training to network. If you can cultivate a business or personal relationship, you can network.

Who should be in your network? Here are some groups to consider:

- Current and former coworkers
- Classmates and alumni of schools and universities you attended
- Members of professional organizations to which you belong
- Relatives, friends, and neighbors
- Members of social, recreational, religious, volunteer, and other groups to which you belong
- Members of community groups and activities

Your network may include people you have never met. You may hear about them through a mutual contact, or you may get their name from a directory or other resource.

Professional networking groups may also be of help. For a listing of networking and support groups by geographical region, use the helpful Riley Guide (www.rileyguide.com/support.html).

MAKING A CONTACT LIST

Networking can help lead you to unadvertised jobs. To use networking to full advantage, it is important to compile a list of your contacts. Some people keep a list of contacts in a computer database, while others prefer using index cards or another written format for each contact. Some information to maintain on each of your contacts is shown in Table 3-1. As the table suggests, you should rate how well you know each contact and how much each can help in your job search.

Table 3-1 Important Contact Information for Networking

Name of person

Title

Employer

Contact information

Email address

How do you know this contact?

How well do you know this contact? (You may want to rate this on a three-point scale such as 1: Know well, such as friend or family; 2: Know somewhat, such as colleagues; 3: Don't know well or don't know at all.)

Discussion notes/referrals

Follow-up

How valuable for job search? (You may want to rate this on a three-point scale such as:
1: Well connected, with good information and help; 2: Moderately connected, can provide some information and help;
3: Not likely to be of much help.)

Comments

Communicating with Your Contacts

Before discussing how to use your contacts to find a job, it's important to understand the basic guidelines for networking.

◆ Seek advice, information, or feedback rather than solutions or jobs.
◆ Avoid asking for information you can get through personal research.
◆ Ask for information your contact would be comfortable giving to you.
◆ Know what you want, be clear about your objectives, and clearly state them.
◆ Let your contacts know how much you appreciate their help by sending them a thank-you note or letting them know how you used the information they gave you. Also, offer to help people who have helped you.

When you are looking for a job, drawing on your contact list for help is a logical choice. Not that you are going to call people to ask them directly for a job—what you want is to find out who has job openings and the names of additional contacts that may lead to other job openings. In the process, you will gain even more networking contacts.

Most people you call will be happy to help you, but they don't have much time, so make your point quickly and directly. To do this, you need to develop a script, or a sales pitch, ahead of time. Your script should include the following elements:

◆ Your name
◆ How you know this person or who referred you
◆ Your occupation
◆ Your current situation
◆ What type of job you're looking for and what you have to offer
◆ A request for names of people to contact regarding job openings, a request for the contact to get in touch with you if he or she hears of a job opening, or a request for advice

Try to fit your information into a 20- to 30-second message. For example, an actual script may look like this.

> My name is John Lucas. Ted Alvarez gave me your name. Do you have a minute or two to talk? [After hearing "yes."] I've been the Chef at the Makefield Country Club for the past three years, and sales have grown 20%. I would like to get some hotel cooking experience, and Ted told me you might know some people I could contact for a hotel Chef position.

Once you have written a good script that is neither too long nor too short, you need to practice it so it comes across clearly and confidently. You also want it to sound spontaneous and genuine, not memorized.

When you are well prepared, these calls will be much easier than you anticipated. You have nothing to lose by calling; if you don't make the call, you'll never find out if there was good information or a job lead at the other end. If you do call and make it easy for the contact to help, you will most likely be successful. At the worst, you'll be a bit uncomfortable. Each call you make is good practice for the next.

Of course, you can also use email and letters to contact people. Again, you need to include all the elements just discussed, being sure to get to the point in an upbeat manner. Letters usually work best with contacts you know fairly well.

In addition to using your contact list, be sure to attend association meetings, conferences, annual meetings, and so on. These are all excellent places to network with fellow professionals. People are approachable at meetings, so use your script to find out about job openings and obtain more contacts. Also, be sure to bring lots of business cards and copies of your resume so you can distribute them if asked. Keep a record of all the contacts you make, what the results were, and any follow-up that is needed. This will help you organize your time and monitor your progress.

TARGETING EMPLOYERS

Many employers want you to find them. This technique is a little different from networking in that you identify employers for whom you would like to work, and then contact them to see if any jobs are, or will be, available.

Don't confuse this technique with mass mailings of resumes to zillions of potential employers. You want to conduct a targeted mailing, not a mass mailing. First, research a number of potential employers, choose which ones to contact, and write an individualized letter to a specific person at each. As with other techniques, follow up with phone calls.

The best place to start learning about prospective employers is often the Internet. Many employers have websites that are full of information, although they usually just share the good stuff! Look for sections on websites entitled About Us, Pressroom, or News Releases. These sections contain background information and the company's most recent news. In addition, publicly held companies publish annual reports that tell you just about everything you need to know. The president's message in the annual report can say a lot about the mission and future of a company. The annual report is often on the website.

Other good places to learn about employers include networking contacts, trade magazines, and association publications. Local business journals also have information about local businesses including news, names, and upcoming events. For a list of business journals for many U.S. cities, go to www.bizjournals.com/ journals.html. Also, be sure to look in your local Yellow Pages.

After you have compiled a list of interesting employers, you need to find the person who does the hiring. Although it is easy to simply send your resume to Human Resources, that does not always get results. Based on the type of position you want (for example, Banquet Chef), find out who the supervisor is (for example, the Executive Chef) and send him or her your resume. To get the Executive Chef's name, ask an appropriate networking contact or simply call the operation and say, "I'm writing a letter to the Executive Chef and I would like the spelling of the name. Can you help me?" The section on cover letters later in this chapter will help you send an appropriate letter with your resume.

Search the Public Market

The public job market includes jobs that are advertised in some way to potential applicants. This section discusses college placement offices, job fairs, classified advertising, recruitment and executive search firms, and employment agencies. Employer websites also often list job openings. This topic is discussed in the section on using the Internet in your job search.

COLLEGE PLACEMENT OFFICE

The college placement office provides a number of services to students. Many offer workshops on writing resumes and taking interviews, for example. Go to your college placement office often to check on workshops and other resources as well as to become friendly with the counselors, who can be helpful in your job search. Check often on when recruiters are coming and sign up for interviews.

JOB FAIRS

Job fairs, like interviews, are face-to-face meetings between jobseekers and employers. They are among the easiest places to find good job leads. Every employer attending is there to find good candidates for open positions.

At a job fair, jobseekers gather information about a company to help them decide if they want to apply for a job. Recruiters staff booths and answer questions, distribute brochures, accept resumes, and conduct mini-interviews. Here are tips to help you make the most of a job fair:

- ◆ Dress as you would for an interview (see the next chapter).
- ◆ Bring a briefcase, expandable folder, or bag to organize materials such as brochures, application forms, and business cards.
- ◆ Bring copies of your resume.
- ◆ When you arrive, take a quick walk through the fair. Time is limited, and booths are probably crowded. Plan a route to see the companies you really want to visit. Save visits with the best prospects until you've first warmed up with a few other employers.
- ◆ If you are with a friend, don't appear to be inseparable when visiting booths. A professional image is easier to maintain if you speak to employers alone.
- ◆ Be professional during the mini-interview. Introduce yourself, shake hands, answer questions, and ask questions. Find out about the company, the types of jobs available for people with your education, and hiring procedures. Don't leave the booth without getting the recruiter's business card.

◆ After each mini-interview, it is often a good idea to take a few notes so you don't confuse what each recruiter told you.

CLASSIFIED ADs: WHERE TO LOOK

You are certainly at least a little familiar with the want ads, a feature of newspapers across the country. Small hometown papers, major metropolitan dailies, and national newspapers have classified sections that include "Help Wanted" advertisements. Hometown papers tend to offer local jobs, often with small companies, while the bigger metropolitan dailies offer jobs over a much larger geographical area; they may even offer national or international jobs. Some newspapers post their classified advertising online.

Another source of want ads is trade magazines such as *Nation's Restaurant News* and association publications. Look in the back of these publications for classified advertising, including want ads. You may already subscribe to some of these publications or have access to them at a college library.

When using classified ads, keep these tips in mind:

◆ When you start seriously looking at want ads, scan all of them to see which categories and position titles you need to target. Don't limit yourself to looking under C for cooking and Chef jobs. Employers often list cooking or Chef jobs under the industry name: restaurant, hotel, hospital, etc. Get an idea early on where to look for relevant ads.
◆ Read the ads daily. Sunday editions tend to have the largest classified section.
◆ When you are deciding which ads to respond to, try to select only those where you meet all or most of the qualifications, and you are really interested in the job.
◆ Be careful about responding to blind ads—that is, ads that do not mention the name of the employer. Blind ads may be used by employers who don't want anyone to know that a position is, or will be, vacant. Sometimes, though, blind ads are used by companies that have a poor reputation. Treat blind ads carefully; if you are working, the employer might be yours!
◆ Answer ads promptly, within a few days, because openings may be filled even before the ad stops appearing in the paper.
◆ Keep a record of all advertisements to which you respond. You might simply staple the ad to a copy of your cover letter.

RECRUITMENT AND EXECUTIVE SEARCH FIRMS

Recruitment firms, also called search firms or headhunters, work for client companies to find qualified candidates to fill specific positions. In return for finding a candidate who gets hired, the recruiting firm is paid a fee by the client. Recruiters work in many industries, including the foodservice and restaurant arenas. Many recruiting firms specialize in a specific industry. Several Internet websites offer directories of recruiting firms; use these to locate only those companies that recruit for your job.

Executive search firms, as their name indicates, focus on recruiting for top management positions. Both recruitment and executive search firms can give you access to jobs you might not otherwise hear about.

If you want to see if a recruitment or executive search firm has jobs you might be interested in, check the website to see if you can register and if they have any positions posted that could be a match for you. If so, send an email with your resume to the recruiter and then follow up with a telephone call several days later. Recruiters usually know their client companies well and are good at matching candidates to jobs. Other ways to get in touch with a recruiter are to ask someone to introduce you (the preferred way) or to send a strong cover letter with your resume. Keep in mind that the recruitment firm is much more approachable than the executive search firm because the recruitment firm normally is filling midlevel positions or lower, whereas the executive search firm is only filling top management positions.

If you are contacted by a recruiter and you pass the initial phone screening, normally the next step is to meet with him or her. Recruiters interview you carefully to see if your background and personality match the needs of any of their clients. Be prepared to fill out application forms and to have your references checked carefully.

If a recruiter has a position that would be suitable for you, he or she should be your guide throughout the interview process. Good recruiters prepare you for your interviews and provide coaching and feedback. They also help to arrange appointments and travel schedules.

EMPLOYMENT AGENCIES

Employment agencies can be public, meaning they are run by the government, or private.

Government Organizations

In line with the U.S. Department of Labor's vision for helping jobseekers, CareerOneStop (www.careeronestop.org) is a collection of electronic tools that are available because of a federal-state partnership. Each of these tools offers help to jobseekers.

1. **America's Job Bank**—This is the biggest and busiest job market in cyberspace. Jobseekers can search for job openings and post their resume where thousands of employers search every day.

2. **America's Career InfoNet**—You can get millions of employer contacts, wage and employment trends, and state-by-state labor market information here.

3. **America's Service Locator**—This service directs you to one-stop career centers in your area that offer job postings, information about local employers, help preparing for job interviews, posting of your resume, and many other services.

Private Employment Agencies

Private employment agencies can be helpful, but they are in business to make money. Most operate on a commission basis, with the fee dependent on a percentage of the salary paid to a successful applicant. You or the hiring company will pay the fee. Find out the exact cost and who is responsible for paying associated fees before using any service.

Although employment agencies can help you save time and contact employers who otherwise might be difficult to locate, the costs may outweigh the benefits if you are responsible for the fee. Contacting employers directly often generates the same type of leads that private employment agencies provide.

The Internet

The Internet can yield tons of information relevant to your job search. Most of what the Internet has to offer can be categorized into one of these five areas.

1. Advice and counseling on resume writing, interviewing, and other career and job search topics
2. Information on employers and salaries
3. Networking
4. Resume postings
5. Job postings

Job hunting using only the Internet is likely to be a waste of time. The Internet is just one way to search for a job. Use it wisely, or you will be spending a lot of time job searching online with little in return.

Table 3-2 lists some of the more prominent online job sites. Most of these sites offer career advice, resume postings, and job postings. Big job sites, such as Monster, may offer some type of networking program as well. Table 3-3 lists websites that specialize in culinary and foodservice jobs, as well as professional organizations that post jobs. The websites noted are all current as of this writing but may have changed by the time of your search.

ADVANTAGES AND DISADVANTAGES

Using the Internet has advantages and disadvantages. One advantage is that you can work at home (as long as you have a computer with Internet access) at any time you choose. Once you have posted your resume on job sites that offer search-while-you-sleep capabilities, the job site actually contacts you (by email), or the employer, or both, that a match is found. Using appropriate keywords in your

Table 3-2 General Job Boards

Monster Boards   www.monster.com

To search jobs, choose a Keyword, Location, and Job Category. Highlight Restaurant and Foodservice for Job Category. Use any of these keywords: Chef, Sous Chef, Chef/Manager, Executive Sous Chef, Chef/Kitchen Manager, Executive Chef, Corporate Chef, Food Production Manager. You can also post your resume and get job search advice and information.

Flipdog   www.flipdog.com

Flipdog is a Monster company and offers most of the same features.

America's Job Bank   www.careeronestop.org

Sponsored by the U.S. Department of Labor, this site allows you to search through a database of over one million jobs, create and post your resume, and do a job search in your local area for Food and Lodging jobs.

Career Builder   www.careerbuilder.com   Yahoo!   hotjobs.yahoo.com

These sites offer places to find jobs, post your resume, and check out advice and resources for finding a job.

U.S. Office of Personnel Management   www.usajobs.opm.gov

This site offers a place to find jobs, post your resume, and check out advice and resources for finding a job with the federal government. Click on Search Jobs. Enter Cook Supervisor, Chef, Cook.

Table 3-3 Websites for Culinary Jobs

American Culinary Federation	www.acfchefs.org
Chef2Chef Culinary Portal	www.chef2chef.net
Chef Jobs	www.chefjobs.com
Chef Jobs Network	www.chefjobsnetwork.com
Foodservice.com	www.foodservice.com
Gigachef.com	www.gigachef.com
HCareers.com	www.hcareers.com
National Restaurant Association	www.restaurant.org
Star Chefs	www.starchefs.com
Restaurant Report	www.restaurantreport.com/Jobs/Index.html

resume (as discussed in Chapter 1) is crucial to finding matches. Job sites can also often provide salary information—which, however, may be inaccurate—and employer websites contain much useful information, including job postings.

Yet another advantage of the Internet is that websites dedicated to helping people network have been growing. For example, Ryze.com and Linkedin.com help you build a professional network with fellow members. Networking sites not only help you contact people you wouldn't ordinarily meet but also many members are hiring managers from national companies who are looking for a better way to meet qualified candidates than posting a job and receiving thousands of emails.

Although the Internet may sound like a jobseeker's dream, it has its downside. Because so many job hunters are using the Internet, you have a lot more competition when you respond to a job posting. Also, job postings are sometimes out of date, incomplete, or simply too plentiful to search through without a big expenditure of time. Using the Internet is also impersonal. It may be high-tech, but it's also low-touch.

OVERCOMING THE SHORTCOMINGS OF THE INTERNET

So how do you overcome the shortcomings of the Internet? First, be sure to use other job search methods, especially networking (both in person and online). Second, show initiative and persistence by taking a couple of steps beyond just emailing your resume to an employer. After you email your resume to a potential employer, mail a paper copy with a cover letter to the person that same day. Then, within a week, if you have not been contacted, call the person to check if your resume arrived and if you can set up an interview. By this time, you should have done some homework on the company so you can sound interested and knowledgeable on the telephone. Don't forget that job candidates stand out when they make the extra effort to contact and establish a relationship with the hiring manager. Although you may think you are being a nuisance, you're not. You are being graded by how much you show initiative and persistence.

ELECTRONIC RESUMES

You need two versions of your resume when using the Internet. Both versions contain the same words; they just use different formatting. One version is your original word-processed version. Employers sometimes ask that you attach your word-processed resume to your email. The second version is referred to as an ASCII resume or a plain text resume. If you're clueless about ASCII, don't worry. ASCII is just a form of computer file that is easily understood by many kinds of computer programs. In short, it's a stripped-down version of your resume file that doesn't contain special formatting or symbols. It can be used to paste your resume into specified fields on job board resume builders or online job applications. It can also be pasted directly into an email message. To create an ASCII resume, follow these four steps.

1. Reset your margins. Each email software program has its own length of lines that is acceptable. To make your resume easy to read on any software program, your resume should have no more than 65 to 70 characters and spaces per line. Set your page margins at 1 inch for the left margin and 1.75 inches for the right margin.

2. Convert your resume to an ASCII file. Word-processing software, like Microsoft Word, can easily create an ASCII file. Just click Save As (using a different file name) and select Text Only with Line Breaks. If you are an XP user, select Plain Text, and when the File Conversion window appears, click Insert Line Breaks under Options. Then click OK. Your file should now have a.txt extension (the part of the file name after the period).

3. Clean up your resume. Make the following changes to your resume:
   - Change bullets to asterisks or hyphens (dashes).
   - Get rid of columns or tables. (This might be easier to do before step 2.)
   - Delete any references to "Page 2" and multiple appearances of your name.
   - For emphasis, use all capital letters rather than boldface or italics, but do so sparingly. Words with all caps are harder to read.
   - Use the space bar, not the tab key, to space text.
   - Rearrange text as needed. Review every line for extra spaces, words in wrong places, and so on.
   - Leave space between sections to make them stand out.
   - Check for misspellings and grammatical errors. Misspelled keywords will be skipped over by the scanner.

4. Check it out. Once you're finished with the conversion and cleanup of your resume, open it using WordPad, Windows' simple word-processing program. Print your resume and then review it. If you've done your job right, it should look good. Also, paste it into an email to yourself or a friend so you can check it out one more time.

RESPONDING TO JOB POSTINGS

Now that you have a plain text resume to use on the Internet, here are tips for responding to job postings online.

- Put the job code, job title, or number in the subject line of your email.
- Address your cover letter directly to the recruiter.
- Unless you are asked to attach your resume, put your ASCII-formatted resume in the body of the email message. This will get your information in front of the recruiter's eyes. When you attach a resume, it takes more time for the recruiter to get to; also, the file may be rejected by email systems due to virus concerns. However, you can also attach your resume, giving the recruiter a choice.

Some job sites sort resumes by date of submission, with the most recent resumes up front. Renew your resume every two to three weeks to keep it fresh.

Following Up on Contacts

Job candidates stand out when they make the extra effort to contact and establish a relationship with the hiring manager. Although you may think you are being a nuisance, you're not. This is how you can make an impression and stand out from the other candidates.

Filling Out Applications

Many jobs require applicants to complete an application instead of, or sometimes in addition to, submitting a resume. Application forms make it easier for employers to evaluate and compare a group of applicants because the forms ask the same questions. In many cases, it is harder to compare resumes. If an employer asks you to fill out an application, do so graciously. Don't bother offering your resume in place of the application. If an employer uses application forms, you must fill one out to be considered.

When given an application form, read it over completely before you begin. Use your resume to help you fill in the necessary information. Write neatly in black or blue ink. Answer every question on the application. Write "none" or "not applicable" if a question does not apply to you.

Although applications do not offer the same flexibility as a resume, you can still find ways to highlight your best qualifications. For example, you can use strong action verbs to describe your job duties and accomplishments. If you do not have paid experience, you can list volunteer job titles.

Applications often ask for your salary history, and your application may be considered incomplete without it. If you are unsure of the exact numbers, write in an approximation, Usually, approximations are acceptable.

If possible, make a copy of your completed application. If you go back for an interview, take the copy with you.

Writing Cover Letters

Every resume you send, fax, or email must have its own cover letter. Sending a resume without a cover letter is like starting an interview without shaking hands. The purpose of the cover letter is not simply to say what job you want and repeat what is in your resume. The best cover letter should emphasize:

◆ your enthusiasm and energy
◆ the employer's interest
◆ an impression of competence
◆ you above the competition

Ultimately, you want your cover letter and resume to generate enough excitement to get you called in for an interview.

So how do you write a great cover letter? It's not hard. Just check out these tips:

◆ Every cover letter should have a professional appearance. Use a block or modified block format that fits on one page, looks neat, and contains no errors. In the block format (see Figure 3-1), all text starts at the left-hand margin, except if you want to put your name and contact information centered at the top, as on your resume. In the modified block format (see Figure 3-2), indent the first line of each paragraph five spaces and place the date, "Sincerely," and your name in the middle of the page. Use the same stationery your resume is printed on for your cover letters.

> Whenever possible, send your letter to a specific person rather than to an office. Consider how differently you respond to a letter addressed to "Occupant" and one addressed to you. If you do not know whom to address, call the employer and ask who is hiring for the position. Check that the name you use is spelled correctly and the title is accurate. Pay close attention to the correct use of Mr. or Ms. Use a colon after the name in the salutation, not a comma, as follows.

Dear Mr. Smith:

◆ If you are responding to a want ad, you can skip the salutation line and go right to the opening paragraph. If you absolutely can't get a name, use "Dear Sir/Madam:" as your salutation.

◆ The first paragraph, called the opening paragraph, should tell the employer which job you are applying for and the connection you have to the company. If someone the employer knows suggested you apply, mention that recommendation. If you are responding to an advertisement, refer to it and the source that published it. You can also put a position reference line between the address and the salutation, as in Figure 3-2.

> Your knowledge of the company might give you another opportunity to connect yourself to the job. You could briefly cite a recent success or refer to its excellent reputation for catered events, for example. You might also want to state why you would like to work for the employer. Don't go overboard; save the specifics for the interview.

◆ In the next paragraph, the main paragraph, you highlight your knowledge, skills, abilities, accomplishments, and successes that relate directly to the position for which you are applying. The idea is that the cover letter should complement your resume, not just repeat it. One way to do this is to summarize your most relevant credentials using a bulleted format. Leave no doubt in the reader's mind that you can contribute to the success of the operation.

◆ In your final or closing paragraph, thank the reader for his or her time, request an interview, and repeat your home phone number. The closing is your chance to show commitment to the job.

Figure 3-1

Cover Letter Using Block Format with Centered Name and Address

Figures 3-1 and 3-2 show sample cover letters.

Email cover letters are much briefer than typed letters to ensure ease of readability. Addresses are not needed, but be sure to use a salutation. The position you are applying for should be typed into the subject line. As in Figure 3-3, include a bulleted list of reasons you should be considered for the job, along with

---

Heather Plumb, C.W.P.C.

211 West Greenwich Avenue
Greenwich, CT 07041
203-437-9365 (h) 203-530-8821 (c)
brewchef@yahoo.com

May 6, 2009

Mr. Ted Carlisle, CEC
Executive Chef
Wamtuxet Inn
10 Shore Drive
Madison, CT 08483

Dear Mr. Carlisle:

As a Certified Working Pastry Chef, I am looking for a position in a larger operation. Marybeth Gilmore gave me your name because she said you plan to expand your bakery operation soon. I have read many fine reviews of the food at the Wamtuxet Inn and feel I could contribute to its fine reputation.

My ten years of employment in the pastry field show increasing responsibility, dedication, and solid accomplishments, such as the following.

- Developed and executed a new menu for 100-seat bakery operation.
- Increased sales 30% by selling bakery items to area restaurants.
- Reduced employee turnover from 75% to less than 25%.
- Developed new quality-control standards.

Thank you for your time reviewing my enclosed resume, which can only briefly highlight my qualifications. I look forward to an opportunity to meet with you to discuss how my interests and qualifications can best meet your needs. I will call next week to schedule a convenient time for an interview. In the meantime, please feel free to call me at 201-437-9365.

Sincerely,

Enclosure   Heather Plumb, CWPC

211 West Greenwich Avenue
Greenwich, CT07041
203-437-9365 (h) 203-530-8821 (c)
brewchef@yahoo.com

May 6, 2009

Mr. Ted Carlisle, CEC
Executive Chef
Wamtuxet Inn
10 Shore Drive
Madison, CT 08483

Re: Pastry Chef Position

Dear Mr. Carlisle:

As a Certified Working Pastry Chef, I am looking for a position in a larger operation. Marybeth Gilmore gave me your name because she said you are looking for a new Pastry Chef for your expanded operation. I have read many fine reviews of the food at the Wamtuxet Inn and feel I could contribute to its fine reputation.

My ten years of employment in the pastry field show increasing responsibility, dedication, and solid accomplishments, such as the following.

- Developed and executed a new menu for 100-seat bakery operation.
- Increased sales 30% by selling bakery items to area restaurants.
- Reduced employee turnover from 75% to less than 25%.
- Developed new quality-control standards.

Thank you for your time reviewing my enclosed resume, which can only briefly highlight my qualifications. I look forward to an opportunity to meet with you to discuss how my interests and qualifications can best meet your needs. I will call next week to schedule a convenient time for an interview. In the meantime, please feel free to call me at 201-437-9365.

Sincerely,

Heather Plumb, CWPC

Enclosure

Figure 3-2 Cover Letter Using Modified Block Format

a request for an interview and any other required information. Unless you have specific instructions on how to email your resume, paste it below the cover letter in text format and also attach it as a Word document. Microsoft Word is the industry's standard word-processing program.

Subject: Pastry Chef Position

Dear Mr. Carlisle:

My strong qualifications for the available pastry chef position and a referral from Marybeth Gilmore have prompted me to contact you. In addition to being a Certified Working Pastry Chef, I have over ten years of experience in the pastry field and have:
- Developed and executed a new menu for 100-seat bakery operation
- Increased sales 30% by selling bakery items to area restaurants.
- Reduced employee turnover from 75% to less than 25%.
- Developed new quality-control standards.

I would like to meet with you to discuss how I could contribute to the fine reputation of the Wamtuxet Inn. Thank you.

Heather Plumb, CWPC
****

My resume is pasted below in text format, and I have attached a Word copy if you prefer to download it.

Figure 3-3 Email Cover Letter

EXERCISES

1. Write up at least 12 cards on networking contacts. Rate how well you know each contact as well as how much each can help in your job search.

2. Write a short script to use when calling someone about a job. Be sure to mention the contact who referred you.

3. Find out what services your college placement office provides.

4. Find out if there are any job fairs in your local area over the next eight weeks. Use the Internet, classified advertisements, and other sources.

5. Attend a job fair and speak to at least two recruiters. What is different about talking to a recruiter compared to having a formal interview?

6. Attend a meeting or conference of a professional organization such as the local chapter of the American Culinary Federation and make at least two networking contacts.

7. Generate a plain-text (ASCII) resume and bring it to class to compare with another student's.

8. Find an advertisement for a job that interests you. Write a cover letter to the employer. Bring it to class and discuss it with another student.

9. Write a cover letter to be emailed to respond to the job advertisement in exercise 8.

CHAPTER 4

Three-Step Interviewing

Introduction

LANDING AN INTERVIEW IS IMPORTANT. It's like getting your foot in the door. This chapter will help you ace the interview process, get a job offer, and negotiate the details of the job offer. Interviewing is more than going in and answering somebody's questions in an intelligent manner. In an interview, you are selling yourself—your skills, abilities, accomplishments, personality, and more. The length of this chapter alone shows that there's a lot more to interviewing, so start reading! You don't want the interview door to slam shut and leave you without a job offer.

So what is that interviewer thinking about? He or she is looking to see if:

- ◆ You are qualified to do the job (experience, knowledge, skills, and abilities).
- ◆ You would fit in with the company and the people with whom you would be working.
- ◆ You are hardworking, persistent, and passionate about your career.

An interviewer is always looking for a candidate who is a good fit with the job, the supervisor, and the company. Much like every restaurant has its own ambiance, each company has its own atmosphere, and you may very well prefer working for one company over another simply because it matches your style better. An interviewer is also always looking for candidates who talk about what they can do for the employer, not what the employer can do for them.

Most interviews are either with a gatekeeper or the person with the authority to hire you (often the person who will be your boss if you get the job). Gatekeepers are people in human resources departments, employment agencies, or executive search firms who interview you to determine if you should go on to the next interviewing step. They do not determine if you get the job but rather if you should stay in the running. Gatekeeper interviews are also called screening interviews. If you interview satisfactorily with the human resources representative of a managed services company, for example, you will be invited to meet directly with the supervisor of the operation where you would be working. Similarly, if an employment agency or executive search firm interviewer thinks you are a good candidate, he or she will send you to interview directly with the client—your potential employer. In many cases, especially with smaller employers, you interview with the hiring manager from the start. This type of interview is called a selection interview.

Screening interviews may be done by phone. These interviews are becoming more popular because they eliminate the time and money expenses associated with face-to-face interviews. If you get through the phone interview, you will be invited for a face-to-face selection interview.

Even without a screening interview, you may be interviewed several times before a hiring decision is made. With each successive interview, you can expect more technical questions and a closer consideration of how you will fit in. For instance, in your first interview you meet the person who will be your supervisor if you are hired. The interview reveals that you are qualified to do the job. Next, you are invited back to be interviewed by other people on the team. The emphasis in the second interview is not so much on screening you out as on establishing how you will fit in. In addition, the interview covers how you will contribute to the company and be a valuable employee.

Beyond the screening and selection interviews is the confirmation interview. In this third type of interview, the person who will be your supervisor introduces you to his or her boss, usually as a matter of courtesy for approval. In most cases, the superior approves the selection and the supervisor makes the offer. During the confirmation interview, it is important to establish a good rapport with the superior. The issue is not whether or not you are qualified for the job; the superior wants first-hand assurance that you are a great choice. This is not the time to sell yourself too hard. Just be likable and emphasize how you are productive and can meet goals.

Interviews involve three steps. The first step includes everything you need to do before the interview, such as learning about the employer and deciding what to wear. The second step is the interview itself. The final step is what you do after the interview, including sending a thank-you note, evaluating your presentation, and following up.

Before the Interview

1. LEARN ABOUT THE PROSPECTIVE EMPLOYER

Knowing about the employer before you go in for an interview has many positive benefits.

◆ It increases your confidence.

◆ The interviewer looks more favorably on candidates who took the time to research the employer than on candidates who didn't. You appear more knowledgeable, serious, and committed.

◆ It will be easier for you to initiate and follow a conversation about the employer.

◆ It will be easier for you to determine how your knowledge, skills, and abilities can benefit the employer.

The Internet is the best place to begin researching prospective employers. Many employers have websites that are full of information—although they usually just share the good stuff! Look for sections on websites entitled About Us, Pressroom, or News Releases. These sections offer background information and the company's most recent news. Try to find the information noted in Figure 4-1, Interview Form (on page 94). Here are additional resources:

◆ Publicly held companies publish annual reports that tell you just about everything you need to know about a company before an interview. To get a copy of the most recent one, check the employer's website, call the shareholder relations department, or ask a stockbroker. The president's message in the annual report may say a lot about the mission and future of a company.

◆ Other resources for business and financial information on employers include the following:

　◆ **www.ceoexpress.com** — This site has links to lots of newspapers and business periodicals. It also has several search engines.

　◆ **www.hoovers.com** — This site has business information on most American companies.

　◆ **www.sec.com** — This is the site of the Securities and Exchange Commission and has financial information on all public companies.

◆ Your college placement office may have information about the employer. This resource normally maintains files on employers who visit the campus to conduct interviews.

◆ People who work for the employer, or used to, may be good resources. Some of your college teachers may be able to give you the name of such a person. If the person says negative things about the employer, keep in mind that you might have a totally different experience.

◆ Current and past issues of industry periodicals such as *Nation's Restaurant News* may be helpful. You can search the contents of the following industry periodicals at www.findarticles.com: *Food Management, Hotels, Hotel and Motel Management, Nation's Restaurant News,* and *Restaurant Hospitality*.

◆ The employee who schedules your interview may be able to mail you descriptive literature such as company brochures, an annual report, and employee newsletters.

2. PREPARE YOUR QUESTIONS

At some point during the interview, usually toward the end, you will have an opportunity to ask your own questions. This is your chance to find out more about the employer, the job, and who you would be working with. After all, you may have to decide if you want to work there.

It is important to ask questions during your interview. By asking good questions, you show the interviewer you are interested, smart, and confident. Your questions enable the interviewer to see a little more of who you are as well as establish a rapport. Even if the interviewer answers all of your questions in the course of the interview, you should ask at least one when he or she turns the interview over to you. Table 4-1 lists many questions you might ask during your interview.

Table 4-1 Great Questions to Ask the Interviewer

What do you think is the most important contribution the company wants from its employees?

What is the company's mission? (If you found the mission statement on their website, ask the interviewer to discuss it.)

What are the goals of the company for the next five years?

How would you characterize the company's culture? What are its values?

Do you have a job description for this position I can look at?

Why is the position being filled?

What would be my day-to-day responsibilities?

Do you have an organizational chart I can look at? How is the kitchen organized?

What specific skills and abilities are you looking for?

How does this position contribute to company goals?

If I am hired, what will be my first assignment?

What are this job's biggest challenges?

What do you want the person who gets this job to achieve?

What is the budget for this area?

What is my spending authority?

Which committees would I take part in?

How would I be evaluated in this position, and how often?

How will my management and leadership performances be measured? By whom?

Can this job, if done well, lead to other positions in the company? Which ones?

Can you describe the work environment?

What type of employee works here?

What kind of employee is successful here?

How empowered are employees?

What criteria determine who gets this job?

What do you like most about working for this company?

How would you describe your style of management?

How does the company support personal and professional growth?

What training opportunities are available?

Do you have any concerns about my skills, abilities, education, or experience?

Do you need anything else from me to have a complete picture of my qualifications and suitability for this job?

What is the next step in the interview process?

An interview is not the time or place to inquire about salary or benefits. You don't want to seem more interested in financial rewards than in contributing to the company. If asked about salary requirements, try to convey flexibility. The best time to discuss salary and benefits is after you are offered the job. At that point, you are no longer the seller; you are the buyer, and you have more leverage.

3. PREPARE AND REHEARSE YOUR RESPONSES

Another important step in preparing for an interview is to anticipate the questions you will be asked and how you will respond. Pages 86 to 90 show typical questions and responses.

4. CHOOSE WHAT TO WEAR

Dress is not just about receiving respect but also about conveying it. Your appearance at an interview reflects your personal presence in the context of a work culture, and it says a great deal about your work. Remember that the very first contact you have with people is visual.

Make that first impression a good one by taking the right steps to be dressed appropriately. For a cook's position, it is appropriate to wear casual business attire, as described here. For a Sous Chef or higher position, wear professional business dress, as also described below. Do not go to an interview in the Chef's clothes you wore to work that morning. If you are going to take a cooking test as part of the interview and therefore plan to wear your Chef's outfit, make sure everything is perfectly clean and pressed and that your shoes are polished. Here are guidelines for professional and casual business attire:

Professional Business Dress: Men	Professional Business Dress: Women	Casual Business Dress: Men	Casual Business Dress: Women
Suit (navy blue, gray, black)	Suit (navy blue, gray, black)	Dress pants	Dark dress pants
Dress shirt	OR conservative dress	Dress shirt	OR dark skirt with blouse or sweater
Conservative tie	OR dark skirt with blouse or sweater	(jacket and tie not required by highly recommended)	Dress shoes
Dress shoes	Stockings	Dress shoes	Stockings (with skirt)
Dark socks to match shoes	Dress shoes	Dark socks to match shoes	
Matching belt and shoes		Matching belt and shoes	
NO	NO	NO	NO
Loud ties	Miniskirts	Jeans	Jeans
White socks	Very high heels	Shorts	Shorts
Boots	Sandals	Boots	Miniskirts
	Low-cut clothing	T-shirts or polo shirts	Sandals
		Loud ties	Low-cut clothing
			White socks

The objective is to look reliable, not trendy. Don't wear clothes or accessories that draw attention away from you.

Avoid wearing lots of makeup, jewelry, perfume, or cologne, which can be distracting to the interviewer. Make sure your shoes are clean and polished, and check your personal hygiene—hair, fingernails, and so on.

Lastly, avoid last-minute clothing disasters by trying on your interviewing outfit a few days before the interview. Make sure it fits well, looks neat, and is clean and pressed. Also, plan for the unexpected. If you will be wearing stockings, make sure you have at least two pairs. If your shoes have shoelaces, get a spare pair in case they break.

5. CHOOSE WHAT TO BRING

You must take some things with you to the interview, but be sure to pack light. You don't need to lug a huge briefcase stuffed with lots of papers that have nothing to do with your interview. In many cases, a simple writing pad portfolio with a pocket for copies of your resume, references, and Interview Form, plus your calendar (paper or electronic) is enough. If you have a job portfolio in a loose-leaf binder, that's fine too. You must be able to immediately locate papers you want to share with the interviewer and refer to your list of questions. Have your calendar available in case you are asked to schedule another interview. If your cooking skills are going to be tested, bring your own set of knives to be most comfortable. If you carry these things in a slim, professional-looking briefcase, that's fine. Just make sure you have everything ready at least one to two days before the interview, and bring extra copies of your resume and references to hand out.

6. CALM DOWN ALREADY!

Most people are nervous when interviewing—but remember, you were asked to interview for the job because the employer believes you could be right for it. The interview is your chance to confirm that belief and establish rapport.

Also keep in mind that the interviewer is a little nervous too—nervous about selecting the wrong person! Employers often can't obtain a lot of feedback from your past employers, so they rely a lot on the interviewing process, which we all know doesn't always identify the best-qualified person.

To reduce nervousness, get a good night's sleep and maintain your usual routine. You might also call to mind some of your happiest memories or proudest moments before arriving for the interview. These relaxation techniques can also help.

◆ Take a deep, slow breath. Let the air come in through your nose and move deep into your lower stomach. Then breathe out through your mouth. Repeat this for several minutes. Imagine that the air coming in carries peace and calm and that the air going out contains your tension.

◆ Slowly clench your fists. While keeping them clenched, pull your forearms tightly up against your upper arms. While keeping those muscles tense, tense all of the muscles in your legs. While keeping all those tense, clench your jaws and shut your eyes fairly tight. Now, while holding all your muscles tense, take a deep breath and hold it for five seconds. Then, let everything go all at once. Feel yourself letting go of your tensions.

A little bit of nervousness is okay. It will help you think clearly and concentrate.

During the Interview

Now it is finally showtime! Because the interview is the first meeting between you and your prospective employer—and a relatively brief meeting, at that—your interviewer will have to base most decisions about you on first impressions. The manner in which you introduce yourself, your personal appearance, whether you maintain eye-to-eye contact with the interviewer throughout the conversation, the completeness and honesty of your answers to questions, whether you are on time—these factors will combine to form the interviewer's appraisal of you, both as a person and as a prospective employee.

GETTING TO THE INTERVIEW

On the day of the interview, give yourself plenty of time to get ready and travel to the location. Plan to arrive 10 to 15 minutes early, even after allowing yourself extra driving time for traffic jams, roadwork, and other hazards. Consider taking a test drive or testing your public transportation route beforehand. You will be a lot more confident on the day of your interview if you know exactly where you are going.

Once you get there, find a restroom to check your appearance. Make sure to remove your sunglasses, portable stereo, and chewing gum. Use a breath mint if needed. Then check in about five minutes early with the appropriate person. It's important to make a good impression from the moment you enter the reception area. Greet the receptionist cordially and try to appear confident. You never know what influence the receptionist has with your interviewer. If you are asked to fill out an application while you're waiting, be sure to do so completely. If you are instructed to sit down for a few minutes, look over your notes, read through company literature, and go over the major points you want to make in the interview. Keep smiling and be friendly!

General guidelines of what to do and what not to do during interviews are listed in Tables 4-2 and 4-3. Read them over now, before the interview starts!

INTRODUCTORY PHASE

Make a favorable impression at once by smiling and greeting the interviewer by title (Mr. or Ms.) and name, then introducing yourself in a professional, self-confident manner. Never use the interviewer's first name unless you are

Table 4-2 General Interviewing Guidelines

- Turn off your cell phone or pager (or put it in silent mode) before you go into the interview.
- Don't chew gum or candy.
- Maintain good eye contact with the interviewer, especially when he or she is talking. This shows interest and self-confidence. Good eye contact does not mean staring; look away periodically.
- Use good body posture. Stand straight and sit correctly. Do not slouch.
- Show you are open and receptive by keeping your legs uncrossed. Don't cross your arms while you sit; it comes across as being defensive.
- Smile naturally at appropriate times.
- Check that you are not tapping your foot, running your hands through your hair, pulling on your jewelry, or using any other distracting mannerisms that show nervousness.
- Speak clearly and firmly. Don't talk too softly or too fast.
- Answer each question completely and directly. Be concise—most questions can be answered in 30 seconds to two or three minutes.
- Be specific when you answer questions. A good interviewer won't let you get away with being vague. Use specific examples to illustrate your points.
- It's okay to shed some modesty and brag a little about your accomplishments. Just don't overdo it, and don't even think about being arrogant!
- Talk about what you can do for the employer, not what the employer can do for you.
- Be honest.
- Never speak negatively about a former or current employer. It serves none of your purposes and will lower the interviewer's estimation of you. Put yourself in the place of the interviewer. If you speak poorly about one employer, what is to prevent you from speaking poorly about this employer if you get this job?
- Project enthusiasm about the prospect of working in this position.
- Be a good listener. Do not interrupt the interviewer. Instead of anticipating the interviewer's next question, concentrate on each question as it is being asked. Being a good listener is an excellent way to build a rapport with the interviewer.
- Pauses are a normal part of the interview process. It's okay to take a moment to put your thoughts together before answering tough questions.
- If you don't know the answer to a question, it's okay to say so. If you have never run a profit-and-loss account, don't fake it.
- Maintain your self-confidence throughout the interview.
- Let your personality come through.

Table 4-3 What Not to Do During an Interview

- Be late.
- Dress informally.
- Have poor personal hygiene.
- Have bad breath.
- Tap your feet or fingers or click your pen.
- Let the interviewer pose all the questions.
- Be sarcastic.
- Be overbearing.
- Make negative statements about past supervisors or employers.
- Know it all.
- Be overassertive.
- Interrupt the interview constantly.
- Express yourself in an unclear manner.
- Overuse phrases or words such as "I guess," "yeah," and "like."
- Ask too many questions about salary and benefits.
- Ask any of the following questions.
  - Why do I have to fill out this job application when the information is on my resume?
  - Do I get compensation time for hours worked beyond 40 hours a week?
  - Can you tell me about the retirement plan?
  - Can I tape this interview?
  - I missed my lunch. Do you mind if I eat my sandwich while we talk?
  - Will this take long? My girlfriend is waiting for me outside.
  - When will I be eligible for my first vacation?
  - Is it possible to telecommute at all with this job?
  - Would I get an office or a cubicle?
  - Are you single?

invited to do so. Make eye contact and stand up straight. Be ready to shake hands if the interviewer extends a hand. Be sure your handshake is firm, but not firm enough to bruise. Sit down in the seat the interviewer indicates. Sit deep and comfortably in your seat. Take a deep breath. Try not to sit on the edge of the chair and look nervous.

The person conducting the interview will begin to form an opinion of you based on such things as the firmness of your handshake, the clearness of your voice, and whether you walk with purpose or shuffle along, so pay attention to what you are doing. This is not the time to try out the latest slang expressions or to move in low gear.

THE HEART OF THE INTERVIEW

After introductions, the interviewer will probably discuss the company and describe the job. He or she will then ask questions meant to gauge how well you would fill the position. Many employers use resumes as guides, asking for additional details during the interview. In addition to finding out more information, they are observing how well you communicate and interact.

Some jobseekers are so focused on specific answers that they forget to relax and connect with the interviewer. An interview should be conversational, with the normal exchanges and pauses. It's okay to pause—for example, to stop and consider an answer to a difficult or unexpected question.

Certain questions will show up in many of your interviews, so it's a good idea to think about them ahead of time and plan how you will respond. Most of your responses will take from 30 seconds to two to three minutes. Your answers should be concise, but don't be afraid to adequately describe your skills, abilities, and accomplishments. Interviewers want to hear examples of how you use your knowledge, skills, and abilities. Their attitude can be summed up in two words: Show me. This is where your portfolio comes in. Use it to your advantage.

Interviewers recommend rehearsing your answers in front of a mirror or with a friend to gain confidence and poise. You may even want to videotape a mock interview to see how you really look and act. The goal is to become comfortable speaking with an interviewer about your education, experience, skills, abilities, achievement, and goals. Whatever you do, do not memorize your responses. The worst thing is to come across as if you are reading from a script.

"Tell me about yourself."

This is a huge question. Don't make the mistake of giving a huge answer. What you want to do here is sum up your education and experience, then end with a statement about "how my background leads me to your company today to interview for this position." You might even start out with a mention of when you knew you were interested in the culinary field. Here's an example.

> I have been interested in working as a Chef since I worked summers on the New Jersey shore in my uncle's seafood restaurant. I worked in every position in that restaurant but loved being a line cook the most. Once I graduated from high school, I went straight to Middlesex County College to get my culinary degree. While completing my degree, I worked as a line cook at the Auberge, a fine dining restaurant. I learned a lot about à la carte dining, station setup, and mise en place. About five years ago, I got my degree and took a job with the Marriot Hotel in Princeton, where I am today. I've worked all the stations in the kitchen there, including working with banquets and catering, and was promoted to Sous Chef two years ago. I am also an ACF-certified Sous Chef. This is the background that leads me to this interview today.

"What are your strongest points?"

This question is a gift, so use it wisely. Think about your knowledge, skills, abilities, experience, personality, motivation, and so on. Mention four or five strengths, and give a specific, brief example to illustrate each. For example:

> I work well under pressure. For instance, last week the water main outside our building broke, and we had no water. Using our emergency plans, we were still able to feed all our guests satisfactorily until the water was turned back on that night.

"What are your major weaknesses?"

You can take two approaches to this classic question. First, you can mention something that is actually a strength, such as:

- *I'm something of a perfectionist.*
- *I'm a stickler for punctuality.*
- *I'm tenacious.*

Second, you can mention a weakness you can easily overcome, such as:

- *I need more computer training.*
- *I need to learn more about nutritious cooking methods. I've signed up for an online course about it.*
- *I need more experience doing public speaking.*

"What do you hope to be doing five years from now?"

The interviewer is not only looking for information about your ambitions but is also seeing if your expectations for advancement match what the employer can offer. It's okay to want to continue climbing the career ladder; just be reasonable about how long it takes to do so. Here is one possible answer:

> I hope I will still be working here and have increased my level of responsibility based on my performance and abilities.

Avoid citing specific time frames. Talk about what you enjoy doing and about realistic opportunities.

"What do you know about our company? Why do you want to work here?"

This is where your research on the company will come in handy. Describe any encounters you have had with the company and offer positive feedback you have heard from customers or employees.

> Your company is a leader in your field and taking on new accounts every day. You run many of the college foodservices in this area, and

I know, just from some friends, that you're doing wonderfully in your accounts. I would like to work for, and learn from, an industry leader.

You might try to get the interviewer to give you additional information about the company by saying you are interested in learning more about the company objectives. This will help you focus your response on relevant areas.

"What was your greatest accomplishment in your current or last job?"

Give a specific illustration from your previous or current job where the accomplishment was totally your doing and had a positive impact. If you have just graduated from college, try to find some accomplishment from your schoolwork, part-time jobs, or extracurricular activities. Don't exaggerate your achievements, and be sure to mention if others helped.

> When I started my current job, there was no catering menu, so we reinvented the wheel every time we did a catering affair. So I developed a standard catering menu, which our customers have enjoyed using. I have also gotten positive feedback from the cooks on it. It makes their jobs a little bit easier and more predictable.

Use the technique shown in this example when you are asked to describe accomplishments: describe the problem, then the action you took, and the results of your action.

"Why should we hire you?"

Cast your background in light of the company's current needs. Give compelling examples. If you don't have much experience, talk about how your education and training prepared you for the job.

> From our discussion, I think you would agree with me that I have the qualifications and experience to contribute to your company. I am also excited about this position and feel I would fit in well. I am sure I can expand your clientele as I did at my last job.

"Why do you want to leave your current job?"

This is not the time to mention that you can't stand your boss—although that may be true. It is generally expected that if you are looking for a new job you are looking for more money, a bigger challenge, a better shot at advancement, or simply a new environment. Make sure you point out why this job will provide you those things. Never complain, gossip, or whine about a current or past boss as this is not professional behavior.

> I want to develop my potential. I have never worked in a hotel food-service and would like to get some experience doing catering and banquets. Also, this operation is a lot bigger than the one I left.

"Tell me about a problem you had in your last job and how you resolved it."

The employer wants to assess your analytical skills and see if you are a team player. Focus on the solution. Select a problem from your last job and explain how you solved it.

"Describe a time you failed."

This may not sound like a question you want to hear, but you can use it to your advantage. We have all had times when our ideas didn't work. Think of a situation when you goofed, but the mistake didn't cause major problems and you learned a valuable lesson.

> One day I forgot to tell Maintenance about problems the baker was having with the floor mixer. The baker was furious at me the next morning because the mixer was still not working right. I ran to Maintenance and luckily got someone who fixed the problem temporarily until a regular repair person came in. Now I know the value of preventive maintenance and have put all the baker's equipment on a preventive maintenance schedule with our Maintenance department. I also learned to keep a pad and pen in my pocket at all times to write things down.

"How would you approach this job?"

This would be much easier to answer once you are in the job for a few weeks, but you're not that lucky! The interviewer wants to get an idea of what types of actions you will take and if those actions are appropriate. Mention that you will need time to observe and survey the operation before you take action. Name a couple of ideas you might implement after learning enough to do so.

> First, I would like to get to know the people in the kitchen and observe them. From what you have said, it seems that the room service area needs immediate attention. After seeing the menu and how the food is prepared and delivered, and after talking with the employees, I am sure to come up with solutions to the timeliness problem.

"Describe your management style."

This question probes how you work with people. Are you a participative manager? Do you like to empower or delegate to employees? Which do you value more: people or production? In the best situation, you have an idea of how this company treats their employees and whether or not your style matches it.

> My employees would say that I am a very participative manager. I try hard to listen because they are on the front line every day taking care of our guests. You've got to take good care of your employees to keep your turnover low and guest satisfaction high.

"What is your philosophy of cooking?"

Here you need to describe the guiding principles that drive you and your cooking, including your philosophy of foods and cooking, your work ethic, management philosophy, and so on. Your cooking philosophy may be, in brief, to emphasize local, organic foods in simple meals, or to blend traditional with contemporary cooking. In a healthcare setting, your cooking philosophy may be to provide home-style, attractive meals that patients enjoy.

You may find the interviewer asking questions that are not job-related. It is inappropriate for an interviewer to ask about your age, race, religion, or marital status. What can you do if you are asked such a question? Take a moment to evaluate the situation and respond in a way that is comfortable for you. For example, if you are asked about your age, be succinct and try to move the conversation back to an examination of your skills and abilities. Or you might say, "I'm in my forties, and I have a wealth of experience that would be an asset to your company." If you are not sure you want to answer the question, ask for a clarification of how it relates to your qualifications for the job. You may decide to answer if the explanation is reasonable. If you feel there is no justification for the question, you might say that you do not see the relationship between the question and your qualifications for the job and you prefer not to answer it.

DON'T FORGET TO ASK YOUR QUESTIONS

Make sure to ask your list of questions. Just as the interviewer is evaluating you, you need to evaluate the job and the employer.

As the interviewer answers your questions, you may want to write down key points. Be sure to ask the interviewer ahead of time for permission to take notes. Asking permission shows that you are polite and respectful. You can phrase the question this way: "Do you mind if I jot down some notes about our discussion? Taking notes helps me organize all the wonderful information I am learning about your company and this job."

ASK FOR A TOUR OF THE OPERATION

Most interviewers will want you to see the foodservice operation. We have prepared a list of points for you to investigate when you take your tour. Not all the items in Table 4-4 are set in stone, but we feel that a "yes" answer to most or all of these points indicates a high-quality place of employment. Keep in mind that, contingent on the type of operation, there are many ways to deal with most of these items. We have listed them to make you think about your potential workplace and how it is managed. After all, you may be spending more waking time there than anywhere else, so the place should live up to your expectations.

WHEN AND HOW TO DISCUSS MONEY AND BENEFITS

The right time to discuss money depends on whether you are applying for an hourly job or a salaried job. For an hourly job, it is appropriate to bring up the

Table 4-4 What to Look for During Your Tour of the Operation

- Uniforms clean and in good condition.
- Adequate and safe locker and changing facilities.
- Kitchen clean and orderly. No standing water, burned-out lights, or accumulated grease.
- Food production areas neat and orderly.
- Garbage area in good order.
- Staff has hats.
- Plastic gloves present.
- Break area in good shape. (This may be part of the restaurant in stand-alone facilities; that is acceptable as long as the management recognizes that breaks and nourishment are part of your workday.)
- Corners, walls, and ceilings clean.
- Refrigeration temperatures correct and temperature logs present.
- Employees happy and well directed.
- Facility maintained and in good repair.
- The kitchen stocked with needed equipment and smallwares.
- Food products wrapped, dated, labeled, and stored in proper containers.
- Temperature of the kitchen reasonable and the air fresh.
- General sanitation apparently correct.
- Hoods clean.
- China stored in an organized fashion.
- Chef certified at Chef de Cuisine or higher.
- Chef recently certified in sanitation.
- Chef's shoes clean and polished.
- Cooks taste what they are preparing.

topic during your initial interview. Often you will know the hourly rate before the interview.

For salaried positions such as managerial jobs, it is risky to bring up salary issues during the interviewing process unless the interviewer does. It is best not to discuss your specific compensation package, especially salary, with the employer until you are offered the job, at which point you are in a much better position to discuss and negotiate salary. Remember: He who mentions money first loses.

If an interviewer asks what your salary requirements are, say you have a range that depends on the whole compensation package of salary, bonus, and benefits.

If pushed, have a range in mind from your minimum salary requirement to 15 to 20% above that figure. Keep in mind that employers know you are looking to make more money than your current or last job, so put your minimum salary requirement above, but not outrageously beyond, your current or most recent salary. You also need to have a handle on the going rate in your locale for the type of position you want. Sources of salary information appear in Table 4-5.

Table 4-5 Sources of Salary Information

Internet

jobstar.org — Jobstar.org has lots of salary review information. Click on Salary Info.

www.salary.com — Try the Salary Wizard for salary information.

www.salaryexpert.com — This website can give you local salary information for many culinary jobs.

www.bls.gov — The Bureau of Labor Statistics has tons of salary information; just be sure to check the date. On the homepage for the Bureau of Labor Statistics, look under Occupations and click on Occupational Outlook Handbook. Next, click on A–Z Index. This book can tell you about earnings as well as the nature of the work, working conditions, employment, advancement, and job outlook for many occupations. Look under any of these three categories in the Index:

B — Bakers

C — Chefs, cooks, and food preparation workers

F — Food and beverage serving and related workers; foodservice managers

Make sure you are looking at the most current edition of the *Occupational Outlook Handbook.*

stats.bls.gov/oco/cg/cgindex.htm — This site is the index to the Bureau of Labor Statistics' *Career Guide to Industries*. It contains salary information about jobs in eating and drinking places, hotels and other lodgings, and health services. Some of the jobs covered are Chefs and Head Cooks, Restaurant Cooks, and Foodservice Managers.

www.bls.gov — The Bureau of Labor Statistics posts salary information by state and metropolitan areas. On the home page, look under Wages, Earnings, and Benefits and click on Wages by Area and Occupation. Next, under State Wage Data, click on By State. You will now see a U.S. map. Click on the state you want, then click on the occupation you want, such as Food Preparation and Serving Related Occupations. Then you can click on 35-0000, Food Preparation and Serving Related Occupations, to get state salary data. At the bottom of this page are links to salary information for metropolitan areas in that state.

Trade Associations

Trade Publications

College/University Career Services Office

Your Network

Your Past Experience

CONCLUDE THE INTERVIEW

Be sensitive enough to tell when the interview is over and it is time to leave. The interviewer may make one of the following statements to hint that the interview is coming to a close.

- I think that pretty much covers it.
- We've covered a lot of ground today.
- I really need to wrap this up.

Instead of saying something, the interviewer might look at the clock or an appointment book, or simply start shuffling papers.

Before the interview is over, be sure to find out what the next step will be. Are you to contact the interviewer, or is the interviewer to contact you? How long will it take for the interviewer to reach a decision? Should you contact the interviewer by phone or by email? If another interview is to be scheduled, get the necessary information. It is important to find out how you are supposed to follow up and then to follow the instructions.

Be sure to make your closing statement to the interviewer a positive one. You went into the interview expecting to land this job; it is hoped that you have reason to maintain this attitude throughout the interview. Now you want to leave the interviewer with the same positive feelings about you that you have presented throughout your meeting. In your closing statement, tell the interviewer that:

- You are very interested in the position.
- You are sure you would do the job well.
- You would enjoy working for the employer.

Also, don't forget to thank the interviewer for his or her time as you say good-bye.

After the Interview

EVALUATE THE INTERVIEW

As soon as possible after the interview, use your Interview Form (Figure 4-1) to help you go over it. Make sure to write down the names of people you met, along with their titles and any thoughts you have about them. Perhaps the Director of Human Resources was bossy and curt while the person who would be your supervisor was easy to get along with. Write down additional impressions from your interview such as which questions you answered well or not so well, what was appealing and not appealing about the job, the people, the employer, or the work environment, and so on.

Next, fill in the Follow-up box with your instructions about the next step(s) in the process. Make sure to write down the appropriate names and dates; note the dates in your calendar as well.

Last, fill in the Goals for Next Interview box. Every interview is a little different, and each offers you opportunities to improve your interviewing skills. After you review the interview, set one or two goals for the next. Perhaps you have photographs of plated desserts you missed the chance to show the hiring manager. Maybe you need to work on talking at a slower pace.

CONTACT INFORMATION

Employer Name _____

Address/Phone/Website _____

Contacts (People you know, the interviewer, people you meet during the interview)

Name/Phone #/Title _____

Name/Phone#/Title _____

Name/Phone#/Title _____

FAST FACTS

Headquarters: _____ Public or Private: _____

Number of Units: _____ Location of Units: _____

Number of Employees: _____ Annual Revenue/Sales: _____

Services/Products/Areas of Expertise _____

Interesting Statistics _____

Competitors _____

Company Strengths _____

Company Challenges _____

Figure 4-1 Interview Form

MY INTERVIEW QUESTIONS

1. _____
2. _____
3. _____
4. _____
5. _____
6. _____
7. _____
8. _____

INTERVIEW IMPRESSIONS

_____
_____
_____
_____
_____
_____
_____

FOLLOW-UP

1. _____
2. _____
3. _____

GOALS FOR NEXT INTERVIEW

1. _____
_____
2. _____
_____

SEND A THANK-YOU NOTE

Even after the interview is over, your task is not complete. Secure a good impression by sending a thank-you letter to the interviewer. It is best to send the letter on the same day.

Thank-you letters should be brief—shorter than one page—and may be handwritten or word processed. Their purpose is to express your appreciation for the interviewer's taking the time to see you and to state again your interest in the job. Most thank-you letters have three main paragraphs (see Figure 4-2).

Figure 4-2
Sample
Thank-You
Letter

1. The first paragraph is your chance to thank the interviewer again for meeting with you and to show enthusiasm for the job. Refresh the interviewer's memory by mentioning the date of the interview and the position for which you applied.

15 Spring Road
Hamlet, LS 41112
561-848-9487

April 15, 2009

Mr. Thomas Atkins
Executive Chef
Hilton Hotels
455 East Greenbush Avenue
Pittsburgh, PA 18944

Dear Mr. Atkins:

Thank you for the opportunity to interview with you yesterday afternoon. I am very interested in the Sous Chef position you described.

My culinary education and work experience as a Sous Chef in another local hotel have prepared me well for the open position. I am especially interested in expanding the banquet business, as we discussed. I would welcome the opportunity to contribute to that effort.

I enjoyed meeting you and your staff and look forward to hearing from you soon. If I can provide any additional information, please call me at 561-848-9487. Thank you again for your time and consideration.

Sincerely,

Peter Gates

2. The second paragraph is for you to briefly repeat the skills that make you well suited for the job. You might also note a topic from the interview that was especially interesting to you. Include any important information you forgot to mention during the interview.

3. The third paragraph is where you thank the interviewer again, give your phone number, and state that you look forward to hearing from him or her.

Write or type the letter on solid white, off-white, or gray stationery. Use a standard business format. Put a colon after the interviewer's name and a space after each paragraph. And don't forget to sign your first and last name.

Many employers say an emailed thank-you letter is acceptable if email correspondence was exchanged between the interviewer and the candidate. Otherwise, an email message is not a substitute for standard mail in most situations.

Be sure to proofread the letter, and make sure you spell the interviewer's name correctly. If a group interviewed you, write to each person on the panel or to the person who led and coordinated the interview, mentioning the other people you met. Interviewers tell tales of misspelled, misused words written in thank-you letters that wreck the image of an otherwise impressive candidate. As you write your thank-you note, remind yourself that you might be writing to your next supervisor.

Follow-up

Follow-up is crucial to your success. Job candidates stand out when they make the extra effort to reiterate their strong interest in a position and the company. Although you may think you are being a nuisance, you're not. You are being graded on initiative and persistence. This is how you can make an impression and stand out from other candidates. Contact the employer in the manner you were instructed to—phone, email, or in person. Repeat your interest in the job and ask if you were successful in obtaining a job offer.

Employment Testing

Some employers use tests or other assessment tools as part of their screening process. This section discusses tests you may encounter.

COOKING TESTS

Many of today's kitchens require practical cooking tests. Surprisingly for many, this can be the most intimidating part of the interviewing process. Most of the formats are truly not difficult in themselves; the pressure comes from what's riding on your performance. In most cases, the evaluators will be looking for skills relevant to the position you are seeking. For example, prep cook applicants may be tested on their ability to produce vegetable cuts, while a Sous Chef may be asked to prepare finished dishes. A common practice for the levels above cook is to ask the candidate to prepare a meal from what is called a market or mystery basket. In our opinion, the market basket is far too intimidating to the average Chef and, more important, not an effective evaluation tool. We heartily suggest that all candidates be fully informed as to the foods available or even asked to order the goods they want to use, with plenty of advance notice so candidates may plan their cooking. This scenario allows the employer to evaluate the candidate's planning and organizational skills as well as cooking skills.

Regardless of the venue or format, as cooks we all should have the same fundamental approach to any given task as well as the same basic concerns about accomplishing the work at hand. Those concerns are as follows.

You as a Cook

Just as you wear a suit to the interview, you wear a perfectly clean, well-fitting uniform to the cook test. Your tools are spotless, your shoes shine, and your manners are impeccable. It is not foolish to buy a new uniform especially for this occasion. The uniform can reflect your personal style to some extent, but unless you really know the style of the Chef, keep it reasonably traditional. Understated style is rarely incorrect. Make sure your knives are sharp. Also, make sure to include gloves in your toolkit as well as special spices that may not be available. Tuck into your pockets some Band-Aids so you don't waste time running around looking for one if you need it.

Think about the foods you are most comfortable with, the dishes you know best and you know will work. This is not the time to experiment or try something new to show your creativity. Do now what you do best.

When you arrive, have a list of questions to ask so you are not constantly stopping and asking, "Where is this?" and "Do you have that?" Before beginning the cook's test, ask if you can tour the kitchen to familiarize yourself with the area. (As a side note, be sure not to smoke at any time while in this establishment. They can find out you smoke after you are hired.) You should set up your station, follow an organized game plan, and work quietly with a calm, controlled intensity, all the while taking time to chat a bit with the people around you. Most kitchens these days do want a cook to be personable and in good spirits.

Sanitation

In most, if not all modern kitchens, sanitation and clean work habits are on the top of the list. You are expected to be certified in sanitation, or at the very least able to complete your tasks without violating the local sanitation code. Your workstation should be neat and clean at all times, and you should display an understanding of the importance of why we as cooks are concerned with this subject at every turn.

Basic Skills

Every level of cooking calls for a skill set that applicants are expected to have. At each higher career level, that skill set expands. This applies to cooking methods as well as knife cuts and fundamental product processes such as roux, beurre manié, and the other starches used to give texture and visual appeal to every sauce. You should be able to demonstrate the skills commensurate with the position and type of establishment to which you are applying.

Organization

A cook is basically a creature of habit and is therefore expected to be able to work in an organized fashion. This applies, for example, to managing small demonstration trays of cuts, the waste of various vegetables manicured for use in a three-course meal. Prepare food as if you were in your own kitchen. Separate garbage from trim that can be utilized later. For example, don't throw out chicken trimmings if you would normally use them in stock.

Flavor Development

If the test is for a lower-level position and no finished dishes will be prepared, flavor development is not a factor. However, when dishes are to be formulated, executed, and eaten, you'd better believe they need to taste good! The unfortunate truth is that all too often cooks are so overcome with all the other aspects of the test that they forget their reason for being—that is, to apply heat to edible substances to make them palatable and tasty. I would, in all cases, make this my first order of business when deciding what type of dish to make and how to prepare it. Do not forget all the wonderful herbs, spices, liquids, and the other flavor enhancers available to us. And don't be afraid to use ingredients such as salt, butter, cream, and oil in your cooking, as long as you execute to standard. When used correctly, these treasured ingredients do enhance food flavor.

Craftsmanship

Craftsmanship reflects the level at which you aspire to be employed and should grow with your tenure as a Chef. By no means should a Chef ever be without craftsmanship. This is what the skills that reside in our hands and head do to the food: the way we cut a vegetable, the smooth removal of the flesh from a fish, the single motion of peeling a shrimp, and the slicing of a perfectly cooked roast. These processes require much practice and repetition in order to be done proficiently.

Visual Appeal

Simplicity is the buzzword here. It is the way a cook lays a chicken breast next to a bundle of green beans, neatly drizzles a sauce on half of the breast and part of the plate, then tumbles three nicely browned potatoes into a group, and finishes the whole with a bit of caramelized mushrooms and shallots. This is the quiet skill that reflects a cook's ability to place food on a plate in an appetizing fashion. Don't sell the difficulty short; few culinarians can do this readily without a great deal of thought. Remember that everyone eats with their eyes first.

Nutritional Understanding

Our basic responsibility is to nourish our patrons, so we need to have a prudent understanding of nutrients and how they affect the body. We need to know the good and bad of the foods we wish to serve. This is not to say we must cook in a low-calorie style or must direct our customers to what they ought to eat. However, it is our responsibility to use enhancing substances (such as salt) moderately, meaning to a level that is necessary and not obsessive. For instance, infuse more flavor into a product before you go and add salt.

Culinary Integrity

All things in nature have many uses. The cook's challenge is to discover what each edible substance does best. At the same time, the cook must work to uncover how best to manicure, dismantle, or otherwise ready the product for the cooking process. This involves making the correct and best use of all the secondary parts, which too many of our colleagues irresponsibly refer to as garbage. Don't throw out usable by-products. As examples, you can make bread from the inside of a zucchini, braise hearts of celery, use the insides of tomatoes in sauces and then strain the seeds out, add flavor to stock with chicken trimmings, use meat trimmings for stews, and add artichoke trimmings to stock for a great soup.

Interpersonal Skills

A kitchen is generally hot and busy, and you are going to spend a great deal of time next to your coworkers in high-stress situations. You don't get a cubicle. You need to rely on your coworkers and they on you, so wherever you are working, act as though you are there forever with your coworkers at your side. Treat each person with respect, and almost inevitably it will be returned. Even if it is not, you will never be the worse for the effort. You cannot be a good cook without the rest of the kitchen brigade.

INTEGRITY TESTS

To help ensure they hire honest employees, employers may administer integrity tests. These usually involve two types of questions. The first type is about illegal or dishonest behaviors you may have exhibited in the past; for example, you might be asked if you have ever walked out of a restaurant without paying the bill. The second type asks about your attitudes toward dishonest behavior; for example, you could be asked about your views on punishing shoplifters. On an integrity test, you also might be asked questions about past involvement with drugs or alcohol.

MEDICAL EXAMINATIONS

Medical exams are given to determine whether you have a physical condition that would prevent you from performing the job. It is illegal to give a preemployment physical exam or to ask about disabilities on the application. Physical exams, however, may be given after the job offer is made. The Americans with Disabilities Act (ADA) gives people with disabilities rights that prevent them from being unjustly rejected for a job. If you have a disability or medical condition that you think may pose barriers to your being hired, your state Vocational Rehabilitation Agency can offer assistance.

DRUG TESTS

Drug tests indicate the presence of illegal drugs. An increasing number of companies use drug tests to screen candidates for all job categories, including managers and professionals. You should be aware that some medications, and even some foods, can produce a positive reading even though you have used no illegal drugs. It is important to inform the employer of any such medications you have taken recently. Be aware that drug tests may not be completely accurate. If you are told that your sample indicated drug use but you know you haven't used any illegal substances, ask if a formal appeals process is available. Tell the employer you would like to take the test again. Perhaps you can ask if there is a more sophisticated test you can take.

Tips for Special Situations

SCREENING INTERVIEWS WITH HUMAN RESOURCES

If your first interview is with a human resources representative, keep these few thoughts in mind. First, human resources interviewers can't offer you the job you want but they can reject your application, so you must be as attentive in this interview as in any other. Next, the interviewer is not likely to know everything about the job you're interested in but can tell you lots about the employer and work environment. He or she is checking your qualifications against a list, so make sure you present yourself well. The human resources interviewer also may ask questions about inconsistencies on your resume, such as work history gaps, and questions to reveal what kind of person you are. Last, be sure to make the human resources interviewer your ally.

Here are additional tips for interviewing with human resources representatives.

- Be very interested in the employer and the job.
- Be upbeat and positive.
- Be confident—but not overconfident.
- Treat the interviewer with respect.
- Listen carefully to what the interviewer says.
- Tell the interviewer why you are qualified to do the job. Give an example or two to back up the information in your resume.
- Keep your answers straightforward.

- Ask the interviewer about his or her experiences working for this employer and what attracted him or her to this employer.
- Gather information about the position and who you would be working with.

Keep in mind that human resources interviewers are often the most skilled interviewers you will meet. They are good at finding out if you should go on to the next interview.

TELEPHONE INTERVIEWS

As mentioned, the telephone is often used for screening interviews to trim time and travel expenses. The procedure for this type of interview is the same as for face-to-face interviews: You need to prepare, and you need to follow up. Although a phone interview sounds informal, don't be caught off guard. You have to succeed in this interview to get to the next step, and you won't have body language to help you interpret how the interview is going. Your only clue is from the voice on the telephone.

When you are scheduling this interview, ask the employer for the names of the interviewers. More than one person may be involved. Make sure you know their names in advance; this makes it easier to identify who is talking during the interview.

Here are tips for telephone interviews:

- Use a phone in a quiet location where the interviewer won't hear distracting background noises. Avoid using a cell phone because the sound is better on land lines and you won't have to worry about the battery dying!
- Imagine that the other person is in the room and talk directly to him or her.
- Your tone says a lot, so try to sound animated and enthusiastic. It helps if you smile, too!
- Have a copy of your resume on hand to help you answer questions and give you confidence.
- Have paper and pen handy.
- Take the opportunity at the close of the interview to persuade the interviewer that you are the ideal candidate for the position.
- Don't rush—it's not your phone bill!
- Some people find it helpful to wear interview clothes so they can focus and project better.
- At the end, say thank you. Don't forget to send a thank-you note!

PANEL INTERVIEWS

You may be asked to interview with a panel of two or more people who each have an interest in who fills the position. For example, you may be interviewed by the person who will be your supervisor, his or her supervisor, a human resources representative, and one or two potential coworkers. Panels are commonly used to interview candidates for teaching positions. The so-called search committee comprises mostly teachers who interview the candidates together.

Here are tips for handling panel interviews:

◆ Greet each person by name. Once you sit down, try to make a quick seating chart to help you remember each panel member's name.
◆ Each interviewer comes to the meeting with a different point of view and set of questions for you. Don't expect the interviewers to have a uniform front; they won't. Treat each one as an individual with his or her own quirks and questions.
◆ The interviewer who talks the most may not have the most say about whether you get the job. Conversely, the interviewer who says the least may have the most say. You don't know, so ignore no one and treat everyone respectfully.
◆ As in other interviews, ask questions when appropriate.
◆ Maintain eye contact with all panel members.
◆ At the end of the interview, thank the group for having you.

If You Get a Job Offer

If you get a phone call extending an offer, what do you do? Do not immediately accept, even if this is your dream job. It is perfectly acceptable to ask for one day to consider the offer, including the compensation package. However, do not negotiate salary or anything else until you have had time to think over the offer. In any case, it is better to negotiate in person than on the phone.

DO YOU WANT THIS JOB?

For each job you're offered, list the pros and cons and evaluate the offers using the following criteria.

◆ Specifics about the position: Duties, position level, hours of work, working conditions, travel requirements, and so on.
◆ Potential for growth and promotion; time frame for performance evaluations and rate increases.
◆ Salary and benefits:
   ◆ Starting salary
   ◆ Overtime/compensation time policy
   ◆ Bonus plans
   ◆ Vacation policy
   ◆ Sick day policy
   ◆ Personal day policy
   ◆ Holiday schedule
   ◆ Health insurance, including employee contribution
   ◆ Dental insurance
   ◆ Life insurance

- ◆ Retirement savings plans
- ◆ College tuition reimbursement plans
- ◆ Stock options
- ◆ Training programs
- ◆ Provision and laundering of uniforms
- ◆ Moving expense benefit (if applicable)
- ◆ The company: Growth, success, reputation, management
- ◆ Your supervisors: Qualifications of Chef and Sous Chef, personalities, interactions, expectations, and so on
- ◆ Turnover rate
- ◆ Location: Housing availability and costs, recreation, quality of schools, and so on
- ◆ Any other factors you consider important

So how does this job rate? Only you can answer that question.

There will be times when you receive a job offer that is perfect . . . except for one thing. Rather than turning down the offer, consider negotiating with the employer.

NEGOTIATION STRATEGIES

Negotiation is a nonadversarial type of communication in which two parties work together to come to an acceptable agreement. Only serious issues based on realistic expectations should be negotiated. Negotiable items include salary and benefits such as vacation time, sick leave, health insurance, life insurance, and tuition reimbursement. Your base salary and performance-based raises are probably the most negotiable parts of your compensation package.

Be aware that many companies offer a cafeteria of benefits whereby you select from a number of benefit options based on a total monetary value. In other words, the company spends a certain amount of money on each employee for benefits, and employees have some flexibility in selecting how those dollars are spent. For example, employees with children might select childcare reimbursement benefits, while employees interested in going back to school might choose tuition reimbursement.

When negotiating your compensation package, it is important to keep in mind the total package. Make sure to consider all the benefits the company has to offer, not just salary. Before you begin negotiating your compensation, decide which benefits are most important to you so you are ready to talk to the employer.

Here are a few things to keep in mind when entering into negotiations with an employer:

- ◆ Most employers expect to renegotiate some aspect of your compensation package.
- ◆ Negotiate only after an offer is made. Remember, He who mentions money first loses.
- ◆ Negotiate salary before benefits. If you can't get the salary you want, you might be able to make up for some of it with benefits.

- Realize that your current earnings usually provide the starting point for salary negotiations.
- Figure out the absolute lowest salary you will accept.
- Find out the employer's salary range for the job.
- Know what you are worth. Be familiar with salary ranges and typical benefits in the area where the job is located. See Table 4-5 for help in finding regional salary information.
- Present a realistic salary range that demonstrates your knowledge of the local market.
- Anticipate objections the employer might be able to raise (such as other employees at the same level earning less or lack of budget) and be prepared to justify your cost effectiveness.
- Negotiate based on your qualifications, skills, and experience. Demonstrate the benefit to the employer in paying you more.
- Ask questions such as "Your offer seems a bit modest. What would it take to get to a higher level within the pay scale?"
- Don't bring your personal needs to the discussion.
- Don't overnegotiate.
- Be a good listener. Make eye contact.
- Make the negotiation a friendly experience, because if you decide to accept the offer, the person you're negotiating with is likely to be your boss.
- Act professionally throughout the negotiations.

If you can't get your requested salary, ask for a performance review within a certain number of months so you will be eligible for an increase sooner. Alternatively, ask for a one-time sign-on bonus. Work with the employer to look at other avenues.

EXERCISES

1. How would you handle yourself differently in each of the following types of interviews: screening, selection, and confirmation?

2. The following website offers an excellent tutorial and tips on how to research companies online; use the website to do the following assignment: http://www.learnwebskills.com/company/index.html

   Pick an employer for whom you might one day want to work and use the resources noted in this chapter to find out at least six of these points of information: location of headquarters, public or private, number of units, number of employees, annual revenue, history, competitors, strengths, challenges.

3. Find the annual report for a managed service company. What are the major sections of the report? Identify five pieces of information in the report that would be useful to know for a job interview.

4. What would you wear to a job interview with a professional dress code? A casual dress code? Would you need to buy new clothes?

5. Develop responses to five of the questions interviewers like to use.

6. Describe how you would start an interview and how you would close an interview.

7. Describe interviews you have had. How are interviews different for hourly versus salaried positions? Have you ever been offered a job you didn't take? Why did you decline the offer?

8. For a salaried position, what should you do if the interviewer asks how much you want to earn?

9. Use the sources in Table 4-5 to get an idea of how much restaurant Chefs make in your area.

10. Much of the interview process revolves around whether you are the right person to fill the job. We also emphasized in this chapter that the interview is your opportunity to evaluate the employer. What questions can you ask and what actions can you take to see if this is a place you would like to work?

11. How would you handle yourself in each of these interviews: screening interview with a human resources representative, telephone interview, and panel interview?

12. Go to the following website, which contains links to many excellent resources on interviewing; read one of the resources and write a paragraph on what you learned: http://www.careerinfonet.org/crl/library.aspx and type "interview" into the keyword search.

13. Plan a cook test menu and timetable.

PROFILE

L. Timothy Ryan

To read the journeys from industry trail blazers will be an inspiration in following their impressive work ethics, dedication, and perseverance. Reading about them will help you create your own legacy. You are at a crucial time in your career to excel from your youthful energy while solidifying your foundation, learning skills that will be invaluable for your future. These industry leaders are not only our colleagues but also advisors from whom we continue to listen and learn. They were born to lead, teach, and mentor. Read on, and you'll know exactly what we are saying.

L. TIMOTHY RYAN, EdD, CMC, President, The Culinary Institute of America

Q/ Your current position is President of the CIA, which sounds like a big responsibility as well as a lot of fun. Can you give me a brief description of what your job encompasses?

A: Well, it is a lot of fun. The CIA has been in existence since 1946. We have alumni all over the United States and all over the world, and 2,400 degree-seeking students. We also educate about 10,000 continuing education students a year in each professional culinary field, and we do that here at Hyde Park and our campus at Napa Valley, which is called Greystone.

Q / Can you give me an idea of where your career started and what was the path you chose to stay on task?

A: I was born and raised in Pittsburgh, Pennsylvania, so that's where I started. By chance, when I was 12,

I started to work in a local restaurant as a dishwasher. The way that worked out was some kids in my neighborhood had jobs washing dishes. They were older than me, and they wanted to go to a ballgame on a Saturday night. They knew that they couldn't go if they didn't have somebody to cover their shift. So they came to me and convinced me to do it even though I told them I didn't know what to do. They said the Chef was cool and you got paid ten bucks or something like that. So I did it, and actually it was neat, and the Chef was cool. It seems that many of us who are Chefs started out washing dishes. That's a real common denominator.

I came from a poor family, and we were lucky if we went to a restaurant once a year, so I wasn't exposed to restaurants and Chefs. I mean, this was the farthest thing from my mind.

Here I was in this kitchen. For a 12-year-old boy, I got to run this cool machine and squirt stuff. Through the stacks where the shelves were, I had a complete view of the kitchen. There were these folks in white jackets, checkered pants, hats, and knives moving around; it was like a ballet. The Chef was a big, tall, handsome guy who was Italian. You could see that everybody respected this guy. He was very charismatic. It was fun washing dishes and doing all those things. I got to watch all this stuff, be exposed to things I had never seen before. At the end of the night, he took this huge wad of bills out of his pocket, more money than I had ever seen at one time. He peeled off a ten-dollar bill and gave it to me. He told me I did a good job and asked me if I wanted to have a steak dinner, which was another big deal. Then he asked me if I would come back. You bet I did; that is how it all started.

Prior to that, I thought I wanted to be a lawyer, but I didn't know any lawyers. My family didn't know any lawyers. We were poor folks from the inner city of Pittsburgh. My only exposure to a lawyer was from television. I kept coming back to this restaurant. I was exposed to a different kind of role model. He was respected, talented, and rich. He lived in a mansion and owned a couple of restaurants. So now here was a person that I knew and was working for, and could see, listen, watch, and talk to, so I switched my mind. Okay, I didn't want to be a lawyer anymore because that was just something on TV. That was great by my family, so all that worked out well for me and I continued to wash dishes and pots and loved it.

Shortly thereafter, I was watching the cooks and knew I could do that. That's what I really wanted to do, and I asked the Chef if I could cook. And he said, "No, you don't want to do that." I said, "Yeah, I do." "No, I don't think so," said the Chef. "You stay in school and get an education. What did you want to be before this?" he asked. "Well, I thought I wanted to be a lawyer." "You be a lawyer." I was really sort of crestfallen. He said this is a tough industry and you may not understand that, when everybody else is playing, that's when you work. "I'm not sure you're really ready for a lifestyle like that. Why don't you be a lawyer?" the Chef said. So, like most kids, as soon as you hear "no," you want it ten times more.

So I kept pestering him, and he finally let me start to do some basic things. In reflecting back, those were some of the fondest memories of my times in the kitchen because I knew nothing, so everything was a new adventure. Many Chefs can relate to when the callous on your fingers where the knife hits gets bloody and sore, and how proud you were of that callous developing because it meant you were starting to become more professional. I would write things down in my little notebook. When I learned how to make hollandaise sauce or all the different things that I learned how to make, I loved it. He was a great guy and great mentor; he really did things the right way and instilled some lessons that I keep to this day about how to cook, how to take care of the customer, and how to work. I always draw back on those things.

At the same time, as I went into high school, I thought I also did have a natural sort of affinity for the work. I had been inclined to go to college, and it was an expectation to go to college. So really, not knowing any better, I said I'll just go to college for that. I walked to the local Carnegie library, which was about three blocks from my house, and I started to look for colleges that I could go to learn how to be a Chef. And there was only one—located in New Haven, Connecticut, and that was the Culinary Institute of America. So I remember that very vividly discovering in book after book, in reading about the CIA, that this place existed for me.

Q / And that was 1974–1975?

A: *I was here from 1975 to 1977. Then I went out in the restaurant business and got to travel the world, spent time in France and Switzerland. I went back to the restaurant business in Pittsburgh, my hometown. That's where I met Ferdinand Metz, my predecessor at the CIA. Incidentally, I'd never intended to go back to my hometown, but two months into my studies, my father passed away. I had two younger sisters, so I was compelled to come back to my hometown. Then I just ended up staying there. I had already envisioned that I would go to work in New York, but that didn't work out. That was just my destiny. Sometimes that's the way life works out. I was in the restaurant business and doing very, very well, and made a nice reputation for myself in Pittsburgh, and that's where I met Ferdinand Metz.*

Q / Where was he in Pittsburgh?

A: *He was in charge of research and development at H. J. Heinz Company, which has their world headquarters in Pittsburgh. I met Ferdinand and then, shortly thereafter, he became president of the school. He had been to the restaurant several times. I was the first person that he really hired because he was looking to do some different things at the school. He brought me here to help develop and open the American Bounty Restaurant 20 years ago.*

Q / **Twenty years ago, there was no American cooking.**

A: Right. If you talked about American food, people said that is hot dogs and hamburgers. So, the American Bounty restaurant concept was a revolutionary idea.

Q / **We've always had American food and the regionality of America; it just needed to be brought into a restaurant.**

A: That's why I came to the CIA. I really didn't want to teach. That wasn't part of my plan. Chefs of my generation, I think in particular, were dominated by the French. The French were the best Chefs in the world, the most famous Chefs in the world. So I had a French restaurant, that was my inclination. I often tell our students here today that Paul Bocuse was, in our business, the equivalent of Elvis. He was the Beatles. Chefs of my age aspired to be Bocuse.

But when Ferdinand came along, I wanted to talk about this American restaurant. I did some research on it, but I wasn't that excited about it. I thought he was a cool guy, and I certainly had never envisioned that I would come back to my alma mater, though I loved it. Particularly at this point in my career, it wasn't where I was at. I wanted to be the Bocuse of America, just as Dean Ferring, Bradley Ogden, and Jeremy Towers did, and that was in the back of my mind. Ferdinand was very talented and well respected, so I figured I'd come here and do a little stint for two to three years and then be back out on my uncharted course.

But that's not the way that it worked out. The reason why is because I really found what I believe to be a higher calling once I got here. As much as I love being a Chef in the restaurant business, as a faculty member, I feel that I'm changing people's lives. And I'm going to make a far bigger impact here, even if I had become the American equivalent of Paul Bocuse. I'll be touching people's lives in a real and important way, helping them fulfill their dreams and helping them shape the profession and change the industry. That's what kept me here for all these years, from my faculty position, to about six administrative jobs, each with different adventures and different responsibilities. I did well, and that is how I ended up as President.

Q / **I want to talk to you about education, which is incredibly important in this day and age. You said it yourself: In 1977 you went out, you worked, got a little education, and went back out in the field. In 18 months you trained to be a Chef. That's not the case anymore. The foundation of education is as crucial as the foundation of culinary skills. Because this particular book is for new culinarians starting out, I would like your thoughts on this. It was very cool to be a dishwasher and work 100 hours a week to learn. Basically those days are over, and the educational part of this industry is so important.**

A: I think, and I know you agree with this, that you learn a lot at the dish table in an establishment. I can still, sitting here right now, feel what it's like to stand at my dish machine, smell what it was like to be there, and remember the interactions I saw between people when they were good and when they totally blew it. I remember many occasions when I saw things going on in jobs, even when I was a cook. I said if I ever have a chance to be the boss, I would never do that. And so those are some of the most valuable lessons. This is an industry like many professions where you still have to pay your dues. You can't rocket to the top until you learn all those important lessons, earn your stripes, and know what it's like to be a dishwasher. Then, when you're supervising dishwashers in the future, they know you know what it's like. And that's important in our industry.

So I still believe in hands-on. But one of the things that I think is important for anybody who wants to get into this business, or any line of work, is to have a tremendous thirst for knowledge and desire to constantly get better. So that thirst for knowledge led me in later years to continue to formalize my education and go back and get a bachelor's degree and then an MBA and finally to get a doctoral degree. But, along the way, I always read to stay informed. That's the way I approach things. When I picked up golf, I read every book, trying to learn as much as I can. So . . . my inclination was go to a library and see how much I could learn about being a Chef, about cooking and the industry. I had the desire for knowledge coupled with the desire to constantly get better. Even as a dishwasher, one of the things that made me stand out is I didn't just mindlessly do the job. I was trying to think about how I could do this better and faster. I would go to the Chef and say I think if we moved these shelves over here, it'll be better for the waitstaff because we won't drop as many plates on the floor. I remember him saying, "Nobody ever asked that before. I think you're right. Let's move them." I remember how empowering that was to me. Hey, they took my idea. I think having a thirst for knowledge coupled with a desire to constantly improve can propel you to the top of whatever part of the industry you choose.

I do think that formal education is becoming increasingly important. I shouldn't say formal education. Let me clarify it: education beyond the education of our craft—beyond the culinary part. The world is so much more complex. Business is so much more complex. We are dealing with legal issues, financial issues, union issues, human resource issues, and marketing issues. The dish table and stove and culinary classes in and of themselves don't necessarily prepare you for that. You have to have those skills, but you need to know something about finance, marketing, operations management, and so on. The easiest way to do that, of course, is through practical experience, but if you have schooling before the practical experience, you can avoid a lot of mistakes. Today, the school of hard knocks doesn't exist in the way that it used to, because money is too tight, business is too competitive, and people aren't going to pay you to learn on the job and make all these mistakes the way that we once could. We have to provide for that kind of education. It's not just business education either that's important. The components of traditional liberal arts education are important, as well, because it gives you a broader perspective of the world. Chefs can't be isolated and focused only on food. You become a better person, a better leader, and I think a better Chef when you know something about art, literature, and languages, for example.

Q/ Languages are huge.

A: Right. If you're a Chef today, you need at least a rudimentary understanding of Spanish, the first language in the kitchen. You know one of the issues in the industry now is the downsizing. Not all hotels, for instance, have the layers of management as in years past. A Chef today must be an astute business person, not only managing costs but an influential part of budgets and forecasts. It is difficult to learn on the job. Number one, the work ethic doesn't allow the 12- to 15-hour days anymore. Number two, the Chef doesn't have time to take you under their wing to explain the whole financial process, which really is an academic part of education.

Q/ What were the investments you made in the last ten years that really advanced your career?

A: Well, with respect to communication, in high school I took two courses on public speaking. I don't remember the tangible lessons that we learned, but what I do remember is that they forced me to get up in front of a group and make a speech. Once I did that, I realized I was actually pretty good at it and stopped being afraid. From the time I was a high school student, I felt that I could get up and communicate with people. I've continued to build on it.

I made huge investments of my time, money, and effort to better educate myself. There are two thirsts that I find are within me: professional and educational. On the professional side, I was involved in many competitions as well as the culinary Olympics. When I was a student I did not envision myself doing that. They showed us an Olympic film, and I thought it was exciting. I've been fortunate through my career to meet people along the way who encouraged me. I was too naïve at the time to disagree.

For example, I was in Pittsburgh and doing well as a Chef, and there was a man named Roland Schaeffer who was on the 1980 Olympic team. Roland said, "You should try out for the Olympic team." I said, "You got to be kidding. I've never competed in a show." He said, "There's one coming up in Pittsburgh. Why don't you—even if you haven't competed? What's great in the Olympics is the quality of Chefs and you're a good Chef, so why don't you try?" I said, "Roland, I really don't know how." He said, "Well, you do know, and if you need tips, give me a call." I thought about it. Roland set up a little demo, which was maybe 40 minutes, showing how to aspic food for a show. Very naïvely I said, okay, I can do that with no fear.

With no fear, I competed in the food show fully expecting that I would get a gold medal and I did. So my naïveté was fortified and supported. On the basis of my two competitions, I agreed to try out for the Olympic team and basically was off and running on what was a great adventure for me. I constantly try to stress to the students here who are interested in competitions like the Olympics this old Zen proverb, "It's the journey that's the reward."

I know you're a seasoned competitor, Lisa, and you have all kinds of medals and diplomas, 99% of which you don't know where they are. The first couple you get are very rewarding. But then they sort of get lost along the way and, on reflection, it's everything you learn and the people that you meet along the way. So, if I look at the Olympic experience, there are lifelong relationships that I have developed as well as I learned so much about how to be a Chef. I learned to be innovative, which is one of the challenges of the Olympic team. They set new industry standards.

The very top Chefs, the big names in the country and the people who are famous and are making money, they distinguish themselves by coming up with original ideas and, by and large, everybody else copies them. So being part of the Olympic effort taught me how to be creative in a sort of pressure-filled environment. We had practices each month and, if you came to practice without good ideas or with some copied version of somebody else's stuff, your teammates let you know it. I think that that was a good experience and huge investment in time and effort. Then I prepared for the Master Chef test, which is a huge investment but a tremendous benefit.

On the other side, my educational side, there were two people who encouraged me to pursue education. One was Ferdinand. Very early on, I said to Ferdinand, "I'd like to be President some day. What do you think? When you're done, how about me? I could do it." That's not exactly how the conversation went, but he said, if you want to do that, you better go back to school because, in the future, the President must have an MBA; Ferdinand has an MBA. He said the president must have a master's and you will probably need to go beyond that. I said, "Well, I only have an associate's degree, I don't even have a bachelor's degree." He told me to start right away plugging at it. He also went to night school to get his degrees. So I agreed.

There was another gentleman whom you may know, Herman Zaccarelli. He was in charge of continuing education at Purdue University, and very well known in our industry. He was a big supporter of mine and really encouraged me to pursue higher education. He's retired and living in Florida now. He was really a very inspirational person for me. I was working here and went to the University of New Haven in Connecticut to get my bachelor's degree. Maybe took two classes a week. So I would drive down for two hours after work and sit in class for three hours or whatever it was, drive two hours back. I was on the Olympic team as well. I was doing all these other things. I wasn't married, so that made it a little bit easier. I was a single person just focusing on my career. But I just kept going step after step and I eventually, after years, checked off my bachelor's degree and then said, okay, well, I guess I need to work on an MBA and did that. And then I looked at a doctoral degree. And I'm glad to have finished that.

Q/ The 1988 Olympics was a national highlight with a phenomenal team. Winning gold medals in both hot and cold food is an incredible accomplishment while getting your Master Chef certification, which is very difficult. Then continuing your education, bachelor's, master's, and doctoral degree at a very young age is again quite amazing. In recapping these accomplishments, so future leaders can see how one can obtain so much through a career, how do you motivate and encourage your students?

A: Let me tell you just one other story, and it has to do with another point that really impressed me and shaped my future. My predecessor was a great mentor to me, and at one time, we had very much a sort of Yoda-Luke Skywalker relationship, then that sort of grew into more of a partnership over the years. But, when Ferdinand first came to the CIA, he called me, and I was in the restaurant kitchen in Pittsburgh. He said, "I want to talk to you about coming up here." And I was so stunned even that he called me. He had just come back from the 1980 Olympics. I said, "I know you just got back from the Olympics. That's so great. You guys did so well. Congratulations!" This was before he said I want to talk to you about coming up to the school because I was just trying to buy some time and catch my breath for the pure fact he even called me. I was being so complimentary about the Olympics that he sort of nicely cut me off after a little bit and he said, "Yes, we're happy too, but that's in the past and I want to talk to you about the future." I mean, they literally had just gotten back and had done better than any other American team had up to that point. It struck me that he was not even stopping to rest on his laurels for one minute. But you always have to be looking forward, looking ahead at what your next challenge is. Individuals—people fundamentally require challenges. Without a challenge, we shrivel up and die. That's something I believe in.

Q/ Young adults today are more savvy. Like you say, you went to restaurants once a year. Children today go to restaurants once a week, 90% of them. They watch cooking shows. They're online. They're moving and shaking. They listen to music, have the TV going, work online, and do their homework all at the same time. What are your work ethics here at the school to prepare students for the real-world profession and the need to produce leaders, not just cooks?

A: Well, as you said, the students today are different than when we went to school. We're thinking about that all the time here because we have to approach it in different ways. The students will shape the industry in a different way than we did. I think that one of the things they'll force us to do as an industry is to become a little bit more user-friendly in terms of the workforce and to take a different approach to hours. Our generation worked 18 straight hours, no problem. That's not a way to have a life. So they're going to be creative enough to reform things, which will ultimately be better for the industry. It will attract more people that are smart and creative. That's the way that we're going to progress.

Then how are we trying to educate the leaders? One is through higher levels of education, just like we talked about; it's not only about cooking. If you're going to learn to golf, you have to hit golf balls. If you're going to learn to cook, you have to cook food. The only way that happens is in the kitchen. You can't do it out of a book, though academic studies are important. You can't do it watching videos. You have to take a knife in hand and roll up your sleeves. That's the way you build those skills.

In addition, we educate leaders by conveying excellence. We're trying to instill a spirit of excellence, to constantly think how to make things better and how to be a professional.

Q / You've been a big fan of professional organizations.

A: Absolutely. The American Culinary Federation (ACF) was a big part of my development. If you look at leadership development, I was a public relations chair for the ACF for many years, and then I became Vice President, President, and Chairman of the Board. I learned a lot of things about interaction, communication, and leadership. I still promote organizations like ACF or the Research Chefs or the National Restaurant Association. Those are great times when leaders are getting together. In formal sessions, you learn a lot, and in the informal sessions, just talking to your colleagues is one of the most powerful things that you can do. I still continue to support all the professional associations; I think they're critical.

Q / When I went to culinary school, I was going to be a Chef. Years later I realized that I didn't want to continue to be a Chef; I wanted to get into management. I wanted to run the operation, directing the entire team and preparing financials. This industry is so wide open. Can you talk about what you envision in nontraditional career paths such as research and development, manufacturing, business and industry, healthcare, or teaching? Have you seen this industry change in the diversity of available professions?

A: Absolutely. I think that there are more opportunities for young people in this business now than ever before. There's a marketing adage that says, "Over time, all markets fragment." If you take a look at our profession, when you and I went to school, we thought of being a Chef. And to us that meant being a person in a singular, independent restaurant. This is still an option today, but there is a whole world of new openings. As the world becomes more sophisticated and Chefs become more sophisticated, they see more opportunities. So, when we were younger, you looked to a Paul Bocuse or Andre Soltner in America. They were icons that had single restaurants. Our generation of Chefs said I can do well in a single restaurant, but there's all these other business opportunities. Why don't I have ten restaurants, and I'll be much more financially successful. So the top Chefs in the country nowadays have multiple restaurants. One example would be, of course, Thomas Keller. He's had the singular restaurant, the French Laundry. Now he's going to open his second restaurant, Per Se, in New York. And Charlie Trotter did the same. There's too many opportunities out there, and Chefs are becoming businesspeople just like anybody else. Chefs who worked in restaurant kitchens somehow got involved in research and development and said, I can have a completely different kind of lifestyle now. So I think our profession is responding to all the opportunities out there, probably taking a more savvy approach and a more businesslike approach to those opportunities than we did in the past.

CHAPTER 5

Advancing Your Career

Introduction

ADVANCEMENT OPPORTUNITIES FOR COOKS AND CHEFS depend on their training, work experience, ability to perform increasingly more responsible and sophisticated tasks, leadership and management skills, and level and type of education. Large establishments and managed service companies usually offer excellent advancement opportunities. Chefs and cooks who demonstrate an eagerness to learn new cooking skills and to accept greater responsibility can move up within the kitchen and take on responsibility for food purchasing, menu development, and supervision. Others may advance by moving from one kitchen or restaurant to another.

During the early part of your career, it's a good idea to stay in each job at least one year, or as long as it takes to learn everything you can there. If you are no longer challenged, perhaps it's time to look for another job to get experience with a different type of cuisine, work with a new chef, or simply work in another foodservice segment.

As Chefs improve their culinary skills, their opportunities for professional recognition and higher earnings increase. Chefs may advance to Executive Chef positions and oversee several kitchens within a foodservice operation, open their own restaurants as Chef-proprietors, or move into training positions as teachers or educators. Your career path is your own unique creation. This chapter discusses ways for you to advance.

Succeeding in a Professional Kitchen

Whenever you start a new job, you receive some type of training. The most common type is on-the-job training, in which more experienced employees guide you. Large establishments and regional offices of nationwide chain or franchise operations increasingly use video and satellite TV training programs to educate newly hired staff. This type of corporate training generally covers the restaurant's history, menu, organizational philosophy, and daily operational standards. Nationwide chains often operate their own school for prospective managers so they can attend training seminars before acquiring additional responsibilities. Whatever type of training you receive, be sure to pay attention, take notes, and ask questions.

Following are additional guidelines for succeeding in a professional kitchen.

◆ Listen carefully to the Chefs and other supervisors.
◆ Be polite and don't use derogatory or demeaning terms.
◆ Get to work on time and be prepared to work long hours.
◆ Work quickly, efficiently, and neatly, always keeping in mind quality, safety, and sanitation.
◆ Ask questions such as "Would you please show me how to . . . ?" You want to learn as much as possible in each kitchen you work.
◆ Keep a positive attitude.
◆ Spend time getting to know your supervisors, coworkers, and subordinates. This effort will most likely make work more enjoyable.
◆ When the Chef is looking for volunteers, step forward.
◆ Be honest.
◆ Have a sense of humor.
◆ Complete tasks on time.
◆ Follow the rules.
◆ Show respect for and support other team members.
◆ Remember that there is no substitute for work experience.
◆ Get the basics down before you experiment and innovate.

In sum, remember that the world is filled with variables and that you must be flexible, accommodating, and, most of all, patient. Hard work and effort mixed with dedication and loyalty is the recipe for success. In our industry, those qualities are not overlooked; they are commended and rewarded.

Setting Career Goals

It is important that you construct your own career growth plan, since only you can decide where you want to go in the culinary field. Although developing this plan is your responsibility, it is helpful to enlist the guidance and assistance of others, such as your supervisor. Your goals should represent what you ultimately hope to accomplish within a given period of time. Most people set short-term goals (covering from three to five years) and long-term goals (such as ten years).

Your goals may be to obtain a specific position or to work in a specific segment of the culinary field. When you set your goals, don't get too hung up on job titles, as they are not always accurate. Concentrate instead on the knowledge, skills, and abilities you use on the job. Set a reasonable time frame indicating when you would like to reach your career goals. However, keep in mind that your career goals should be realistic and attainable ones that are reachable through your ongoing efforts to gain experience and education.

Develop a plan of activities to reach your goal. Think of this plan as a step-by-step statement of the specific activities needed to enhance your education, skills, knowledge, or experience. The activities should be measurable and tailored to achieve your specific career goals. You must be able to recognize when you are working toward your goal and when your goal has been accomplished. Be specific and set dates.

You should be prepared to commit a portion of your own time and effort to accomplishing this plan. Completing your planned work experience and/ or training activities is your responsibility. You'll need to seek help when necessary.

Today's career paths don't necessarily progress straight up the organizational chart as they used to. At times, you may make lateral moves, possibly into different segments of the foodservice industry, instead of going up. Be flexible. You may wind up down a totally different road and love it!

When You Leave a Job

When you leave a job, always do it on good terms. Even if you were not happy in the job, act professionally and give from two to four weeks' notice. The exact amount of notice depends on the level of your position and the employer's policies. You will burn bridges behind you if you leave without giving sufficient notice or if you decide to have it out with your boss or another employee just before you leave. If you leave on poor terms, a lot more people than you think will hear about it, and you may one day be asking one of them for a job.

Before you leave any job, it's a good idea to ask your employer to give you a written document listing your job title and dates of employment. See Figure 5-1 for an employment verification form. This document comes in handy when, for example, the restaurant you worked at goes out of business or is sold by your old boss. Be sure to get the form returned on or before your last day of work. Figure 5-2 is a sample employment verification and recommendation letter.

Date: _____

Dear Sir/Madam,

This will certify that _____

was employed at _____

from _____ to _____.

The job title was _____.

He/she supervised _____ personnel and worked _____ hours/week.

Signature _____

Title _____

Address _____

_____

Phone Number _____

Description of establishment: (full-service restaurant, hospital foodservice, etc.)

_____

_____

Figure 5-1 Employment Verification Form

May 15, 2009

To whom it may concern,

This letter is to confirm that Salome Parker is a full-time Banquet server at the Embassy Suites Hotel in St. Louis. She was hired 2/20/04 in our Food and Beverage operation where she currently takes care of our demanding clients such as American Express and Merrill Lynch. She works in every capacity of the department including American, Buffet service and many times Russian service for our private VIP events. She is an avid bartender and has a well-polished presentation both on the floor and interacting with our guests.

I have had the pleasure of working with Salome since 2/20/04 when I started with the company. She is an outstanding employee who exemplifies a committed work ethic, taking personal pride in her assignments. Please accept my recommendation on her behalf. If I can be of any further assistance feel free to contact me.

Sincerely,

Mary Huston, CEC
Food and Beverage Director

Figure 5-2   Sample Verification/Recommendation Letter

Mentors

Chef apprentice programs use the mentor model to train new Chefs. A mentor is a person who helps someone, usually a subordinate, grow professionally. Mentors serve as role models and teach, guide, coach, and counsel their mentees (the individuals they mentor).

Although the mentee clearly benefits from the mentor relationship, the mentor gains advantages too, including the personal satisfaction of passing on his or her successes and the recognition for being a positive role model. Being a mentor also means being able to continually practice culinary, management, and interpersonal skills. Mentors are true professionals in the sense that they directly help others. Besides giving invaluable training, mentors boost the confidence of mentees and provide guidance about career moves.

Mentoring may be formal or relatively informal. Informal mentoring occurs all the time in the workplace. It happens when one person simply helps someone, usually a subordinate, and a career-helping relationship develops. Formal mentoring is found in structured programs in which a company or organization matches mentors with mentees. A good match is one in which the mentor gets along well with the mentee, has the expertise the mentee desires, and is a good teacher.

Regardless of whether mentoring is formal or informal, the following guidelines will help both parties get the most out of the relationship.

1. Have an initial meeting to discuss the expectations of both the mentor and the mentee.
2. Plan to commit to a partnership for six months to one year and discuss a no-fault termination in which either party can back out for any reason.
3. Identify mentee goals and make an action plan. Accept that these goals may change.
4. Set up how often the mentor and mentee will meet to discuss progress.
5. The mentee must be willing to accept constructive feedback, try new things, and take risks.
6. The mentor will use listening, coaching, guidance, career advising, and other techniques to help the mentee reach his or her goals.
7. Many mentoring relationships continue long past the initial time commitment, especially as the partners often become friends. When the mentor and mentee decide their work together is completed, they should go over the original action plan and discuss the progress and results. They should give each other constructive feedback that may help in future mentoring relationships.

In many cases, culinary professionals have to find their own mentors, often by networking, and be willing to work hard once they find a good mentoring relationship.

Professional Organizations

Being active in professional organizations is key to job advancement and your career. Professional organizations give you lots of opportunities to network with other culinary professionals as well as these additional benefits (which vary by organization):

- Industry magazines and newsletters that keep you up to date on the culinary industry.
- Annual meetings/conferences to upgrade and update your knowledge and skills.
- Educational seminars to upgrade and update your knowledge and skills.
- Leadership opportunities.

- Industry contacts—for example, with representatives of foodservice equipment manufacturers.
- Job announcements.
- Recognition awards.

The Appendix gives detailed information about each organization.

The American Culinary Federation (ACF) is the largest and most prestigious organization dedicated to professional Chefs in the United States today. Their mission is as follows: "It is our goal to make a positive difference for culinarians through education, apprenticeship, and certification, while creating a fraternal bond of respect and integrity among culinarians everywhere."

ACF offers its members many opportunities to keep their knowledge and skills current through its monthly publications (National Culinary Review and Center of the Plate) and through seminars, workshops, national conventions, and regional conferences. ACF sanctions U.S. culinary competitions and oversees international competitions that take place in the United States. ACF accredits culinary programs at the secondary and postsecondary levels. Local chapters of ACF offer members opportunities to network with nearby culinary professionals. Also, ACF members are simultaneously enrolled in the World Association of Chefs' Societies (WACS), which has 72 official chefs associations as members from around the world.

Certifications

Getting certified in the culinary field is a tremendous asset and can help you advance your career. ACF offers the only comprehensive certification program for Chefs in the United States. ACF certification is a valuable credential awarded to Cooks, Chefs, Pastry Cooks, and Pastry Chefs after a rigorous evaluation of industry experience, professional education, and detailed testing. Table 5-1 describes the certifications ACF offers. Table 5-2 lists other certifications in the culinary/foodservice field.

If you intend to take a certification test, give yourself at least several months to prepare for it so you can pass on your first try. Once you are certified, you will need to meet continuing education requirements to maintain your certification.

American Academy of Chefs

The American Academy of Chefs (AAC) is the honor society of the American Culinary Federation. AAC recognizes culinary professionals whose contributions positively affect the culinary industry. To apply for membership, a Chef's local

Table 5-1 Certifications of the American Culinary Federation

There are 14 levels of certification, and each requires specific qualifications, in addition to knowledge of culinary nutrition, food safety and sanitation, and culinary supervisory management. It is fundamental to the program that work experience is equivalent to the level of certification.

Cooking Professionals

Certified Culinarian (CC): An entry level culinarian professional within a commercial foodservice operation.

Certified Sous Chef (CSC): A chef who supervises a shift or station(s) in a foodservice operation. A CSC must supervise a minimum of two full-time people in the preparation of food. Job titles that qualify for this designation include sous chef, banquet chef, garde manger, first cook, A.M. sous chef and P.M. sous chef.

Certified Chef de Cuisine (CCC): A chef who is the supervisor in charge of food production in a foodservice operation. This could be a single unit of a multi-unit operation or a free-standing operation. A CCC must supervise a minimum of three full-time people in food production.

Certified Executive Chef (CEC): A chef who is the department head usually responsible for all culinary units in a restaurant, hotel, club, hospital or foodservice establishment, or the owner of a foodservice operation. A CEC must supervise a minimum of five full-time employees and pass a practical exam in front of peers.

Certified Master Chef (CMC): The consummate chef. A CMC possesses the highest degree of professional culinary knowledge, skill and mastery of cooking techniques. A separate application is required, in addition to successfully completing an eight-day testing process judged by peers. Certification as a CEC or CEPC is a prerequisite.

Personal Cooking Professionals

Personal Certified Chef (PCC): A chef with a minimum of four years of professional cooking experience. Also required is a minimum of one full year of employment as a personal chef engaged in all aspects of food preparation and serving, menu planning, marketing, financial management and operational decision making.

Personal Certified Executive Chef (PCEC): A chef with advanced culinary skills and a minimum of six years of professional cooking experience with a minimum of two years as a personal chef. A PCEC is skilled in all aspects of food preparation and serving, menu planning, marketing, financial management and operational decision making.

Baking and Pastry Professionals

Certified Pastry Culinarian (CPC): An entry level culinary professional within a pastry foodservice operation.

Certified Working Pastry Chef (CWPC): A pastry culinarian responsible for a pastry section or a shift within a foodservice operation, with considerable responsibility for preparation and production.

Certified Executive Pastry Chef (CEPC): A pastry chef who is a department head, usually responsible to the executive chef of a food operation or to the management of a pastry specialty firm.

Certified Master Pastry Chef (CMPC): The consummate pastry chef. A CMPC possesses the highest degree of professional culinary knowledge, skill and mastery of cooking techniques as they apply to pastry. A separate application is required, in addition to successfully completing a 10-day testing process judged by peers. Certification as a CEC or CEPC is a prerequisite.

Culinary Administrators

Certified Culinary Administrator (CCA): This is an executive-level chef who is responsible for the administrative functions of running a professional foodservice operation. This culinary professional must demonstrate proficiency in culinary knowledge, human resources, operational management and business planning skills. A CCA must supervise the equivalent of at least 10 full-time employees.

Culinary Educators

Certified Secondary Culinary Educator (CSCE): An advanced-degree culinary professional who is working as an educator in an accredited secondary or vocational institution. A CSCE is responsible for the development, implementation, administration, evaluation and maintenance of a culinary arts or foodservice management curriculum.

Certified Culinary Educator (CCE): An advanced-degree culinary professional who is working as an educator in an accredited postsecondary institution or military training facility. A CCE is responsible for the development, implementation, administration, evaluation and maintenance of a culinary arts or foodservice management curriculum. In addition, a CCE must possess superior culinary experience equivalent to a CCC or CWPC.

Table 5-2 Additional Certifications

American Correctional Food Service Association

Certified Correctional Food Service Professional (CCFP)

Certified Correctional Food Systems Manager (CFSM)

School Nutrition Association

School Foodservice and Nutrition Specialist (SFNS)

Club Managers Association of America

Certified Club Manager (CCM)

Master Club Manager (MCM)

Dietary Managers Association

Certified Dietary Manager (CDM)

Certified Food Protection Professional (CFPP)

Educational Institute of the American Hotel and Lodging Association

Certified Food and Beverage Executive (CFBE)

Certified Hospitality Educator (CHE)

Foodservice Educators Network

Certified Culinary Instructor (CCI)

International Association of Culinary Professionals

Certified Culinary Professional (CCP)

International Food Service Executives Association

Master Certified Food Executive (MCFE)

Certified Food Executive (CFE)

Certified Food Manager (CFM)

Military Hospitality Alliance

Registered Military Culinarian (RMC)

Registered Military Hospitality Manager (RMHM)

National Association of Foodservice Equipment Manufacturers (NAFEM)

Certified Food Service Professional (CFSP)

Table 5-2 Additional Certifications (continued)

National Restaurant Association Educational Foundation	**Research Chefs Association**
Foodservice Management Professional (FMP)	Certified Culinary Scientist (CCS)
	Certified Research Chef (CRC)
National Ice Carving Association	**Retailer's Bakery Association**
Certified Ice Carver	Certified Journey Baker (CJB)
Certified Competition Ice Carver	Certified Baker (CB)
Certified Professional Ice Carver	Certified Decorator (CD)
Certified Master Ice Carver	Certified Bread Baker (CBB)
	Certified Master Baker (CMB)

ACF chapter must request an application from the AAC national office as well as sponsor the applicant. Some of the mandatory requirements for membership consideration in the AAC are as follows.

1. Must be certified as an ACF Master Chef, Master Pastry Chef, Executive Chef, Executive Pastry Chef, or Culinary Educator for not less than two years. Must continue to renew certification.

2. Must be in the culinary profession for not less than 15 years. Ten of the 15 years must have been at the Executive Chef or culinary educator level.

3. Must be a member of ACF at least ten years.

4. Must have attended any combination of four ACF- or ACF-approved regional conferences or national conventions within a ten-year period.

In addition to the mandatory requirements, candidates must complete at least ten additional requirements such as being a chapter president for a full term or being a chair or co-chair of a culinary art salon or hot food competition approved by ACF or WACS.

Lifelong Learning

Lifelong learning, also called continuing education, refers to the continuous development of the skills, knowledge, and understanding essential to maintaining employment as well as to meeting personal needs. Chefs must be aware of the latest in cooking techniques, cuisines, purchasing, and much more. How do you keep up? Just check out any of the following:

◆ Professional associations such as the ACF present many opportunities for continuing education. The ACF and its chapters offer publications, seminars, workshops, online courses, and more.

◆ Many cooking schools and colleges offer continuing education classes and resources such as books and DVDs so working culinary professionals can learn about current practices. Some schools, including the Culinary Institute of America, offer online courses as well (www.ciaproChef.com).

◆ Culinary conferences and conventions always include seminars and workshops.

◆ Culinary industry magazines provide current information on trends, etc.

Lifelong learning is vital if you want to advance in the profession.

EXERCISES

1. Pick out a certification you may want to have one day. Go to the appropriate website and find out what you need to do to obtain that certification. Find out how often you have to recertify (such as every five years) and what is required to recertify.

2. Go to the American Culinary Federation website (www.acfChefs.org) and find out the requirements for membership in the American Academy of Chefs.

3. Find three websites that offer culinary lifelong learning opportunities. Which site is most attractive to you and why?

4. Interview a culinary professional who has experience as either a mentor or a mentee. Was the experience worthwhile? How did it work?

APPENDIX A

Culinary Professional Organizations

AMERICAN CULINARY FEDERATION (ACF)

Who Are They?

The American Culinary Federation (ACF) is the largest and most prestigious organization dedicated to professional Chefs in the United States today. Their mission is as follows.

It is our goal to make a positive difference for culinarians through education, apprenticeship, and certification, while creating a fraternal bond of respect and integrity among culinarians everywhere.

ACF offers its members many opportunities to keep up to date with the latest in knowledge and skills through its journals, seminars, workshops, national conventions, and regional conferences. ACF also sanctions U.S. culinary competitions and oversees international competitions that take place in the United States. ACF accredits culinary programs at the secondary and postsecondary levels. Local chapters of ACF offer members opportunities to network with nearby culinary professionals. As a member of ACF, you are simultaneously enrolled in the World Association of Chefs' Societies (WACS), which represents 54 countries.

ACF offers the only comprehensive certification program for Chefs in the United States. ACF certification is a valuable credential awarded to Cooks and Chefs and Pastry Cooks and Pastry Chefs after a rigorous evaluation of industrial experience, professional education, and detailed testing.

Where Are They?

180 Center Place Way
St. Augustine, FL 32095
800-624-9458
www.acfchefs.org

PUBLICATIONS
Monthly magazine: *The National Culinary Review*
Monthly newsletter: *Center of the Plate*

CERTIFICATIONS
Certified Master Chef (CMC)
Certified Master Pastry Chef (CMPC)
Certified Executive Chef (CEC)
Certified Chef de Cuisine (CCC)
Certified Executive Pastry Chef (CEPC)
Personal Certified Executive Chef (PCEC)
Certified Culinary Administrator (CCA)
Certified Culinary Educator (CCE)
Certified Secondary Culinary Educator (CSCE)
Certified Working Pastry Chef (CWPC)
Certified Sous Chef (CSC)
Personal Certified Chef (PCC)
Certified Pastry Culinarian (CPC)
Certified Culinarian (CC)

AMERICAN DIETETIC ASSOCIATION (ADA)

Who Are They?

The American Dietetic Association (ADA) is the largest and most visible group of professionals in the nutrition field. Most members are Registered Dietitians (RDs). Individuals with the RD credential have specialized education in human anatomy and physiology, chemistry, medical nutrition therapy, foods and food science, the behavioral sciences, and foodservice management. Registered dietitians must complete at least a bachelor's degree at an accredited college or university, a program of college-level dietetics courses accredited by the Commission on Accreditation for Dietetics Education, a supervised practice experience, and a qualifying examination. Continuing education is required to maintain RD status. Most RDs are licensed or certified by the state in which they live. They work in hospitals and other health-care settings, private practice, sales, marketing, research, government, restaurants, fitness, and food companies.

Where Are They?

120 South Riverside Plaza, Suite 2000
Chicago, IL 60606
800-877-1600
www.eatright.org

PUBLICATIONS
Monthly journal: *Journal of the American Dietetic Association*
Bimonthly newsletter: *ADA Times*

REGISTRATIONS
Registered Dietitian (RD)
Registered Dietetic Technician (DTR)

AMERICAN HOTEL & LODGING ASSOCIATION (AH&LA)

EDUCATIONAL INSTITUTE OF THE AMERICAN HOTEL & LODGING ASSOCIATION (EI)

Who Are They?

The American Hotel & Lodging Association (AHLA) is a 93-year-old federation of state hotel lodging associations, representing about 11,000 property members worldwide. AH&LA provides its members with assistance in operations, education, and communications, and lobbying in Washington, DC, to encourage a climate in which hotels prosper.

The Educational Institute of the AH&LA (EI) is the leading source of quality hospitality education, training, and professional certification for hospitality schools and colleges and industries around the world.

Where Are They?

American Hotel & Lodging Association
1201 New York Avenue NW #600
Washington, DC 20005
202-289-3100
www.ahla.com
American Hotel & Lodging Association Educational
Institute
800 North Magnolia Avenue, Suite 1800
Orlando, FL 32803
800-752-4567
American Hotel & Lodging Association Educational
Institute
2113 North High Street
Lansing, MI 48906
517-372-8800
www.ei-ahla.org

PUBLICATION
Monthly magazine: *Lodging*

CERTIFICATION
The Educational Institute offers a number of professional certifications, including Certified Food and Beverage Executive. The food and beverage manager must be an expert at providing quality service and have excellent leadership and organizational skills, technical proficiency, and a commitment to high standards.

Certification lasts five years, during which requirements for recertification must be fulfilled. Certificate holders must maintain a qualifying position within the industry and earn points by carrying out various activities. The categories in which points are awarded include:

> Professional experience
> Professional development activities/seminars
> Industry involvement
> Educational services

EI provides a portfolio to help track professional development activities.

AMERICAN INSTITUTE OF BAKING (AIB)

Who Are They?

The American Institute of Baking (AIB) provides educational and other programs, products, and services to baking and general food production industries around the world. Its members range from international food ingredient and foodservice companies to small single-unit traditional and artisan retail bakeries. AIB is a resource for bakers looking for information and expertise in baking production, experimental baking, cereal science, nutrition, food safety and hygiene, occupational safety, and maintenance engineering.

Where Are They?

1213 Bakers Way
P.O. Box 3999
Manhattan, KS 66505-3999
800-633-5137
www.aibonline.org

PUBLICATIONS
Monthly magazine: *AIB Research Technical Bulletin*
Bimonthly magazine: *AIB Maintenance Engineering Bulletin*
Quarterly newsletter: *Bakers Way*

CERTIFICATIONS
Certified Baker — Bread and Rolls
Certified Baker — Cakes and Sweet Goods
Certified Baker — Cookies and Crackers

AMERICAN INSTITUTE OF WINE & FOOD (AIWF)

Who Are They?

The American Institute of Wine & Food (AIWF) is an organization devoted to improving the appreciation, understanding, and accessibility of food and drink. With over 30 U.S. chapters, its members include restaurateurs, food industry professionals, food educators, nutritionists, chefs, wine professionals, and dedicated food and wine enthusiasts.

Where Are They?

213-37 39th Avenue, Box 216
Bayside, NY 11361
800-274-2493
www.aiwf.org

PUBLICATION
Bimonthly newsletter: *American Wine and Food*

AMERICAN PERSONAL & PRIVATE CHEF ASSOCIATION (APPCA)

Who Are They?

The American Personal & Private Chef Association (APPCA) has been training, supporting, and representing successful personal Chefs since 1995.

Where Are They?

4572 Delaware Street
San Diego, CA 92116
800-644-8389
www.personalchef.com

CERTIFICATIONS (through APPCA and ACF)
Personal Certified Executive Chef (PCEC)
Personal Certified Chef (PCC)

AMERICAN SOCIETY FOR HEALTHCARE FOOD SERVICE ADMINISTRATORS (ASHFSA)

Who Are They?

The American Society for Healthcare Food Service Administrators is an affiliate of the American Hospital Association. Members include food and nutrition service management professionals in hospitals, continuing care retirement communities, nursing homes, and other healthcare facilities. Members include Director of Food and Nutrition Services, Director of Dining Services, café/catering/vending managers, clinical nutrition managers, and dietitians. ASHFSA welcomes food and nutrition service professionals from both independent and contract operations.

Where Are They?

304 West Liberty Street, Suite 201
Louisville, KY 40202
800-620-6422
www.ashfsa.org

PUBLICATION
Quarterly magazine: *Healthcare Food Service Trends*

CERTIFICATIONS/RECOGNITIONS
ASHFSA has a professional recognition program called APEX, which stands for Actions for Professional Excellence. The three levels of achievement are:

Level 1. Accomplished Health Care Foodservice Administrator (AHCFA)
Level 2. Distinguished Health Care Foodservice Administrator (DHCFA)
Level 3. Fellow Health Care Foodservice Administrator (FHCFA)

The ASHFSA Professional Recognition Program is designed to recognize those factors that are indispensable to true professionalism: basic and continuing education, experience, and participation in professional and society activities. Levels of recognition are achieved successively. Each level requires additional education, work experience, and participation in society activities. Members may put the acronym for the level they have accomplished after their name (for example, John Hall, DHCFA).

ASSOCIATION FOR CAREER AND TECHNICAL EDUCATION (ACTE)

Who Are They?

The Association for Career and Technical Education (ACTE) is dedicated to the advancement of education that prepares youth and adults for successful careers. ACTE provides resources to enhance the job performance and satisfaction of its members, increases public awareness of career and technical programs, and works on growth in funding for these programs. Its members include over 30,000 teachers, counselors, and administrators at the middle, secondary, and postsecondary school levels.

Where Are They?

1410 King Street
Alexandria, VA 22314
800-826-9972
www.acteonline.org

PUBLICATIONS
Monthly magazine: *Techniques*
E-mail newsletter: *Career Tech Update*

BCA (formerly BLACK CULINARIAN ALLIANCE)

Who Are They?

BCA is an educational and networking association of hospitality and foodservice professionals founded in 1993.

Where Are They?

55 West 116th Street, Suite 234
New York, NY 10026
800-308-8188
www.blackculinarians.com

BREAD BAKERS GUILD OF AMERICA

Who Are They?

The Bread Bakers Guild of America provides education resources for members and fosters the growth of artisan baking and the production of high-quality bread products. Its members include professional bakers, baking educators, home bakers, vendors, and others who share common goals.

Where Are They?

3203 Maryland Avenue
North Versailles, PA 15137-1629
412-823-2080
www.bbga.org

PUBLICATION
Newsletter: *Bread Bakers Guild*

CLUB MANAGERS ASSOCIATION OF AMERICA (CMAA)

Who Are They?

Members of the Club Managers Association of America (CMAA) manage more than 3,000 country, city, athletic, faculty, yacht, town, and military clubs. CMAA provides its members with professional development programs, networking opportunities, publications, and certification programs.

Where Are They?

1733 King Street
Alexandria, VA 22314-2720
703-739-9500
www.cmaa.org

PUBLICATION
Monthly magazine: *Club Management*

CERTIFICATIONS
Certified Club Manager (CCM)
Master Club Manager (MCM)

CONFRERIE DE LA CHAÎNE DES RÔTISSEURS

Who Are They?

The Confrérie de la Chaîne des Rôtisseurs brings together professional and amateur gastronomes in an organization that celebrates the pleasures of food, wine, and spirits and encourages the development of young professionals. The organization is based on the traditions and standards of the medieval French guild of rôtisseurs, or "meat roasters." Members are in more than 70 countries around the world. The United States has its own chapter.

Where Are They?

Chaîne House
Fairleigh Dickinson University
285 Madison Avenue
Madison, NJ 07940-1099
973-360-9200
www.chaineus.org

PUBLICATION
Three times yearly: *Gastronome*

DIETARY MANAGERS ASSOCIATION (DMA)

Who Are They?

The Dietary Managers Association is a national association with over 15,000 professionals dedicated to the mission of "providing optimum nutritional care through food service management." Dietary managers work in nursing homes and other long-term care facilities, hospitals, schools, correctional facilities, and other settings. Responsibilities may include directing and controlling menu planning, food purchasing, food production and service, financial management, employee hiring and training, supervision, nutritional assessment, and clinical care. Dietary managers who have earned the Certified Dietary Manager (CDM), Certified Food Protection Professional (CFPP) credential are also specially trained in food safety and sanitation. Dietary managers may work as foodservice directors, assistant foodservice directors, supervisors, and in other positions.

Where Are They?

406 Surrey Woods Drive
St. Charles, IL 60174
800-323-1908
www.dmaonline.org

PUBLICATION
Monthly magazine: *Dietary Manager*

CERTIFICATIONS
Certified Dietary Manager (CDM)
Certified Food Protection Professional (CFPP)

FOODSERVICE CONSULTANTS SOCIETY INTERNATIONAL (FCSI)

Who Are They?

The Foodservice Consultants Society International (FCSI) promotes professionalism in foodservice and hospitality consulting and helps its members by supplying networking and educational opportunities, professional recognition, and other services. FCSI has members who work in layout and design, planning, research, training, technology, operations, and management.

Where Are They?

304 West Liberty Street, Suite 201
Louisville, KY 40202
502-583-3783
www.fcsi.org

PUBLICATIONS
Quarterly magazine: *The Consultant*
Monthly e-mail newsletter: *The Forum*

FOODSERVICE EDUCATORS NETWORK INTERNATIONAL (FENI)

Who Are They?

Foodservice Educators Network International (FENI) is a group of foodservice educators who work in high schools and postsecondary programs. FENI works with educators to help them advance their professional growth. A key element of FENI is to facilitate the means for culinary educators to share teaching techniques and other information with colleagues and industry partners to enhance high standards of culinary education.

Where Are They?

20 West Kinzie, 12th Floor
Chicago, IL 60610
312-849-2220
www.feni.org

PUBLICATION
Quarterly magazine: *Chef Educator Today*

CERTIFICATION
Certified Culinary Instructor (CCI)

INSTITUTE OF FOOD TECHNOLOGISTS (IFT)

Who Are They?

The Institute of Food Technologists (IFT) has members who work in food science, food technology, and related professions in industry, academia, and government.

Where Are They?

525 West Van Buren, Suite 1000
Chicago, IL 60607
312-782-8424
www.ift.org

PUBLICATION
Monthly magazine: *Food Technology*

INTERNATIONAL ASSOCIATION OF CULINARY PROFESSIONALS (IACP)

Who Are They?

The International Association of Culinary Professionals (IACP) is a group of approximately 4,000 food professionals from over 35 countries. IACP provides continuing education, networking, and information exchange for its members, who work in culinary education, communication, or the preparation of food and drink. Its mission is to "help its members achieve career success ethically, responsibly, and professionally." Many of its members are cooking school instructors, food writers, cookbook authors, Chefs, and food stylists.

Where Are They?

304 West Liberty Street, Suite 201
Louisville, KY 40202
800-928-4227
www.iacp.com

PUBLICATION
Quarterly magazine: *Food Forum*

CERTIFICATIONS
Certified Culinary Professional (CCP)

INTERNATIONAL CATERERS ASSOCIATION (ICA)

Who Are They?

The International Caterers Association (ICA) includes both off-premises and on-premises caterers from around the world. ICA provides education, mentoring, and other services for professional caterers, and promotes the profession of catering to the public, vendors, and others.

Where Are They?

91 Timberlane Drive
Williamsville, NY 14221
877-422-4221
www.icacater.org

PUBLICATIONS
Bimonthly newsletter: *CommuniCater*
Membership includes complimentary subscriptions to *Catering Magazine*, *Event Solutions*, and *Special Events Magazine*.

CRUISE LINES INTERNATIONAL ASSOCIATION, INC.

Who Are They?

Cruise Lines International Association is the world's largest cruise association and is dedicated to the promotion and growth of the cruise industry. CLIA is composed of 24 of the major cruise lines serving North America.

Where Are They?

2111 Wilson Boulevard, 8th Floor
Arlington, VA 22201
703-522-8463
www.cruising.org

PUBLICATIONS
Monthly e-mail: *Fast Facts*
Quarterly newsletter: *Even Keel*

INTERNATIONAL COUNCIL ON HOTEL, RESTAURANT, AND INSTITUTIONAL EDUCATION (I-CHRIE)

Who Are They?

I-CHRIE is the advocate for schools, colleges, and universities that offer programs in hotel and restaurant management, foodservice management, and culinary arts.

Where Are They?

2613 North Parham Road
Richmond, VA 23294
804-346-4800
www.chrie.org

PUBLICATION
Monthly journal: *Journal of Hospitality and Tourism Education*

INTERNATIONAL FLIGHT SERVICES ASSOCIATION (IFSA)

Who Are They?

The International Flight Services Association (IFSA) serves the needs and interests of airline and railway personnel as well as airline and rail caterers who are responsible for providing passenger foodservice on regularly scheduled travel routes. IFSA's membership is dedicated to the advancement of the art and science of this segment.

Where Are They?

1100 Johnson Ferry Road, Suite 300
Atlanta, GA 30342
404-252-3663
www.ifsanet.com

PUBLICATION
Monthly e-mail newsletter: *Onboard IFSA*

INTERNATIONAL FOOD SERVICE EXECUTIVES ASSOCIATION (IFSEA)

Who Are They?

The International Food Service Executives Association (IFSEA) offers its diverse membership opportunities for personal development, networking, mentoring, and community service. IFSEA's membership includes professionals from many areas of the foodservice industry.

Where Are They?

304 West Liberty Street, Suite 201
Louisville, KY 40202
502-583-3783
www.ifsea.com

PUBLICATION
Magazine: *Hotline*

CERTIFICATIONS
Master Certified Food Executive (MCFE)
Certified Food Executive (CFE)
Certified Food Manager (CFM)

INTERNATIONAL FOODSERVICE MANUFACTURERS ASSOCIATION (IFMA)

Who Are They?

The International Foodservice Manufacturers Association (IFMA) represents manufacturers of foodservice equipment. IFMA provides its members with many services, including training, networking, visibility, and marketing information.

Where Are They?

Two Prudential Plaza
180 North Stetson Avenue, Suite 4400
Chicago, IL 60601
312-540-4400
www.ifmaworld.com

PUBLICATION
Bimonthly magazine: *IFMA World*

LES DAMES D'ESCOFFIER

Who Are They?

Les Dames D'Escoffier is an invitation-only culinary organization of successful women leaders. The organization mentors young women, educates the public about the pleasures of the table, awards scholarships, and supports food-related charities. Members include chefs, restaurateurs, cookbook authors, food journalists and historians, wine professionals, food publicists, culinary educators, and hospitality executives.

Where Are They?

212-867-3929
www.ldei.org

MILITARY HOSPITALITY ALLIANCE (MHA)

Who Are They?

The Military Hospitality Alliance (MHA) is the affiliate of the International Food Service Executives Association that focuses on the needs of the military. Projects include a military culinary competition, Enlisted Aide of the Year awards, and more.

Where Are They?

836 San Bruno Avenue
Henderson, NV 89015
888-234-3732
www.mhaifsea.com

CERTIFICATIONS
Registered Military Culinarian (RMC)
Registered Military Hospitality Manager (RMHM)

NATIONAL ASSOCIATION FOR THE SPECIALTY FOOD TRADE (NASFT)

Who Are They?

The National Association for the Specialty Food Trade (NASFT) is an international organization of domestic and foreign manufacturers, importers, distributors, brokers, retailers, restaurateurs, caterers, and others in the specialty foods business.

Where Are They?

120 Wall Street, 27th Floor
New York, NY 10005
212-482-6440
www.specialtyfood.com

PUBLICATION
Monthly magazine: *Specialty Food*

NATIONAL ASSOCIATION OF COLLEGE & UNIVERSITY FOOD SERVICES (NACUFS)

Who Are They?

The National Association of College & University Food Services (NACUFS) is the trade association for foodservice professionals at nearly 650 institutions of higher education in the United States, Canada, and abroad.

Where Are They?

2525 Jolly Road, Suite 380
Okemos, MI 48864-3680
517-332-2494
Fax: 517-332-8144
www.nacufs.org

PUBLICATION
Quarterly magazine: *Campus Dining Today*

NATIONAL FOOD PROCESSORS ASSOCIATION (NFPA)

Who Are They?

The National Food Processors Association provides food-science and technical expertise for the food-processing industry on topics such as food safety, nutrition, and technical and regulatory matters. NFPA members process and package fruit, vegetables, meat, fish, and other foods.

Where Are They?

1350 I Street NW, Suite 300
Washington, DC 20005
800-355-0983

PUBLICATION
Monthly journal: *NFPA Journal*

NATIONAL ICE CARVING ASSOCIATION (NICA)

Who Are They?

The National Ice Carving Association (NICA) is the only organization in the United States devoted solely to promoting the art of ice sculpture across the country and around the world. With nearly 500 members, NICA sanctions and organizes ice-carving competitions in North America, providing standardized guidelines for judging the quality of ice sculptures in competition.

Where Are They?

P.O. Box 3593
Oak Brook, IL 60522-3593
630-871-8431
www.nica.org

PUBLICATION
Newsletter: *On Ice*

CERTIFICATIONS
Certified Master Ice Carver
Certified Professional Ice Carver
Certified Competition Ice Carver
Certified Ice Carver

NATIONAL RESTAURANT ASSOCIATION (NRA)

THE NATIONAL RESTAURANT ASSOCIATION EDUCATIONAL FOUNDATION (NRAEF)

Who Are They?

The National Restaurant Association (NRA) is the lead-ing business association for the restaurant industry.

Together with the National Restaurant Association Educational Foundation, the Association's mission is to represent, promote, and educate (through reports, pub-lications, research, training materials, networking, etc.) the restaurant and foodservice industry.

The National Restaurant Association Educational Foundation (NRAEF) is the not-for-profit organization dedicated to fulfilling the educational mission of the National Restaurant Association. The NRAEF is the premier provider of education resources, materials, and programs to attract, develop, and retain the indus-try's workforce. Examples of NRAEF programs include ServSafe® food safety certification, ServSafe® Alcohol certification, the ProStart® School-to-Career program, the ManageFirst® program, and the Foodservice Management Professional (FMP) certification program.

Where Are They?

National Restaurant Association
1200 17th Street NW
Washington, DC 20036
202-331-5900
www.restaurant.org
National Restaurant Association Educational Foundation
175 West Jackson Boulevard, Suite 1500
Chicago, IL 60604-2814
800-765-2122
312-715-1010 (in Chicago)

PUBLICATION **(through the National Restaurant Association)**
Monthly online magazine: *Restaurants USA Online*

CERTIFICATION **(through Educational Foundation)**
Foodservice Management Professional (FMP)

NATIONAL SOCIETY FOR HEALTHCARE FOODSERVICE MANAGEMENT (HFM)

Who Are They?

The National Society for Healthcare Foodservice Management (HFM) is a professional association rep-resenting healthcare foodservice operators and their suppliers. HFM accepts only members who operate independent operations and are not contracted. HFM offers advocacy for independent healthcare foodser-vices as well as management tools to decrease costs, increase patient and staff satisfaction, and define suc-cessful operational performance. Members are mostly from hospitals.

Where Are They?

355 Lexington Avenue, 15th Floor
New York, NY 10017
212-297-2166
www.hfm.org

PUBLICATION
Monthly magazine: *Innovator*

NORTH AMERICAN ASSOCIATION OF FOOD EQUIPMENT MANUFACTURERS (NAFEM)

Who Are They?

The North American Association of Food Equipment Manufacturers (NAFEM) represents companies throughout the United States, Canada, and Mexico that manufacture commercial foodservice equipment and supplies.

Where Are They?

161 North Clark Street, Suite 2020
Chicago, IL 60601
312-821-0201
www.nafem.org

PUBLICATION
Quarterly magazine: *NAFEM in Print*

CERTIFICATION
Certified Food Service Professional (CFSP)

RESEARCH CHEFS ASSOCIATION (RCA)

Who Are They?

The Research Chefs Association (RCA) brings together chefs, food scientists, and others who work in food research and development in restaurants, food companies, and many other businesses.

Where Are They?

1100 Johnson Ferry Road, Suite 300
Atlanta, GA 30342
404-252-3663
www.culinology.com

PUBLICATIONS
Magazine: *Culinology*
Quarterly newsletter: *Culinology Currents*

CERTIFICATIONS
Certified Research Chef (CRC)
Certified Culinary Scientist (CCS)

RETAIL BAKERS OF AMERICA (RBA)

Who Are They?

The Retail Bakers of America (RBA) is a trade association of independent retail bakeries. The association works to improve the operations and profitability of its members by offering training, networking, conventions, communications, and meetings.

Where Are They?

14239 Park Center Drive
Laurel, MD 20707-5261
800-638-0924
www.rbanet.com

CERTIFICATIONS
Certified Journey Baker (CJB)
Certified Baker (CB)
Certified Decorator (CD)
Certified Bread Baker (CBB)
Certified Master Baker (CMB)

SCHOOL NUTRITION ASSOCIATION (formerly the American School Food Service Association)

Who Are They?

The School Nutrition Association (SNA) has over 55,000 members involved in some way with the National School Lunch Program. SNA works on making sure all children have access to healthful, tasty school meals and nutrition education. SNA does this by providing education and training, setting standards, and educating members on legislative, industry, nutritional, and other issues.

Where Are They?

700 South Washington Street, Suite 300
Alexandria, VA 22314
703-739-3900
www.schoolnutrition.org

PUBLICATIONS
Monthly magazine: *School Foodservice and Nutrition*
Semiannual journal: *Journal of Child Nutrition and Management*

Credential

School Foodservice and Nutrition Specialist (SFNS)

SOCIÉTÉ CULINAIRE PHILANTHROPIQUE

Who Are They?

The Société Culinaire Philanthropique is the oldest association of Chefs and Cooks in the United States. Founded by a group of French Chefs, its members organize the annual Salon of Culinary Arts in New York City.

Where Are They?

305 East 47th Street, Suite 11B
New York, NY 10017
212-308-0628
www.societeculinaire.com

SOCIETY FOR FOODSERVICE MANAGEMENT (SFM)

Who Are They?

The Society for Foodservice Management (SFM) serves the needs and interests of executives in the on-site foodservice industry (predominantly B&I). SFM provides member interaction, continuing education, and professional development via information and research.

Where Are They?

304 West Liberty Street, Suite 201
Louisville, KY 40202
502-583-3783
www.sfm-online.org

PUBLICATION
Monthly e-mail newsletter: *FastFacts*

UNITED STATES PERSONAL CHEF ASSOCIATION (USPCA)

Who Are They?

The United States Personal Chef Association serves the needs and interests of personal Chefs. USPCA offers training, certification, personal Chef software, and other services.

Where Are They?

610 Quantum Road, NE
Rio Rancho, NM 87124
800-995-2138
www.uspca.com

PUBLICATION
Monthly magazine: *Personal Chef*

CERTIFICATION
Certified Personal Chef (CPC)

WOMEN CHEFS & RESTAURATEURS (WCR)

Who Are They?

The Women Chefs & Restaurateurs (WCR) promote the education and advancement of women in the restaurant industry. Their goals are "exchange, education, enhancement, equality, empowerment, entitlement, environment, and excellence." WCR offers a variety of networking, professional, and support services.

Where Are They?

304 West Liberty Street, Suite 201
Louisville, KY 40202
877-927-7787
www.womenchefs.org

PUBLICATION
Quarterly newsletter: *Entrez!*

WOMEN'S FOODSERVICE FORUM (WFF)

Who Are They?

The Women's Foodservice Forum was founded in 1989 to develop leadership talent and ensure career advancement of executive women in the foodservice industry. WFF currently offers three types of mentor programs and many other resources to help women move into senior-level positions. Its membership comes from restaurant operations, manufacturing, distribution, publishing, and consulting.

Where Are They?

1650 W 82nd Street, Suite 650
Bloomington, MN 55431
866-368-8008
www.womensfoodserviceforum.com

PUBLICATION
Monthly newsletter: *Open Doors*

Index